William Harvey King

Electro-Therapeutics

Or Electricity in its Relation to Medicine and Surgery

William Harvey King

Electro-Therapeutics
Or Electricity in its Relation to Medicine and Surgery

ISBN/EAN: 9783337778880

Printed in Europe, USA, Canada, Australia, Japan

Cover: Foto ©berggeist007 / pixelio.de

More available books at **www.hansebooks.com**

ELECTRO-THERAPEUTICS

OR

ELECTRICITY

IN ITS RELATION TO

MEDICINE AND SURGERY

BY

WILLIAM HARVEY KING, M.D.,

ELECTRO-THERAPEUTIST TO THE HAHNEMANN HOSPITAL, MEMBER OF THE
NEW YORK SOCIETY FOR MEDICO-SCIENTIFIC INVESTIGATION, ETC.

———————

NEW YORK
A. L. CHATTERTON & CO.
78 MAIDEN LANE

LIST OF ILLUSTRATIONS.

TABLE OF CONTENTS.

CHAPTER I.

ELECTRO-PHYSICS.

CHAPTER II.

ELECTRO-PHYSIOLOGY.

CHAPTER III.

CHANGES IN NUTRITION.

CHAPTER IV.

ELECTRO-DIAGNOSIS.

CHAPTER V.

GENERAL THERAPEUTICS.

CHAPTER VI.

SPECIAL THERAPEUTICS.

CHAPTER VII.

GALVANO–CAUTERY.

PREFACE.

THIS book is the outgrowth of lessons written for private instruction, and in its preparation three things have been kept constantly in mind : First, accuracy and reliability so far as possible in a work of this character ; second, briefness and condensation ; third, simplicity and comprehensiveness.

This work may be criticised by the scientist for not being scientific enough, but it is not for the scientist that it is put before the public, but as a text-book for the student and general practitioner, and consequently the author has tried to make it as practical as possible. A complete description has been given of all treatments and operative measures, so any person possessing an ordinary amount of adaptability can make the application ; and, in those cases where it is necessary to attend to the minutest details, it has not only been the author's endeavor to give them, but also to impress their importance upon the mind of the reader.

The works consulted in the preparation of this volume are :

"Handbook of Electro-Therapeutics." Erb.

"Medical and Surgical Electricity." Beard and Rockwell.

"Medical Electricity." De Watteville.

"Elementary Principles of Electro-Therapeutics." Haynes.

"Medical Electricity." Bartholow.

"Electro-Therapeutics and Electro-Surgery." Butler.

"Lectures on Electricity and its Relation to Medicine and Surgery." Rockwell.

"Electricity in Medicine." Ranney.

"A Treatise on Electrolysis." Amory.

"Du Catarrhe par le Galvano-Caustique Chimique." Garrigòn-Desarènes.

Also several periodicals, containing the writings of Apostoli and others, and various works on electro-physics.

In conclusion the author wishes to acknowledge his indebtedness to Wm. H. Bleecker, M.D., and S. H. Bartlett (superintendent of the Waite & Bartlett Manufacturing Company), for assistance rendered.

23 West 53d Street, New York.
March 21, 1889.

ELECTRO-THERAPEUTICS.

CHAPTER I.

ELECTRO-PHYSICS.

Dynamic Electricity, or Electricity in Motion.—Under this head we may properly consider the phenomena of the electric current produced by chemical decomposition. It may be remarked here that of the real nature of the electric current, so-called, we know nothing—the name being merely conventional.

When a plate of zinc and a plate of copper are partially immersed in a vessel containing dilute sulphuric acid, without touching each other, a slight disengagement of gas (hydrogen) results. This is set free at the zinc plate. If the two plates are connected by means of the metal wire, as shown in Fig. 1, a larger quantity of hydrogen is set free, but now at the copper plate instead of the zinc. This new action that takes place on account of the two metals being connected, namely, the increase in the amount of gas given off and its being given off from the copper instead of the zinc plate, is not to be explained by chemical action. If we now examine the wire connecting the copper and zinc plates, we find it possesses several remarkable properties:

First. If the bulb of a sensitive thermometer is placed against the wire, the rise of the column of mercury will indicate the presence of heat and determine the thermal property of the wire. Second. If we suppose the zinc plate to be toward the north and the copper plate toward or facing the south, then the wire we are examining will lie in the magnetic meridian of the earth. If we now suspend a magnetic needle from a fiber of silk in the same plane as the wire and just above it, the north pole of the needle will be deflected to the right hand of the wire, or, geographically speaking, to the

east.   This we may term the magnetic property of the wire.
Third.   If a wire connected with a sensitive galvanometer be
brought suddenly close to and parallel to the wire connecting
the zinc and copper, a deflection of the needle will be produced.   This we may term the inductive property of the wire.
We may now state briefly that, when the wire exhibits these
phenomena, an electric current is said to flow or be flowing
through the wire, and the direction is always from the positive
along the wire to the negative pole.   In the dilute acid in
which the plates of zinc and copper are immersed, the flow is
from the zinc, which is the soluble plate, through the fluid to

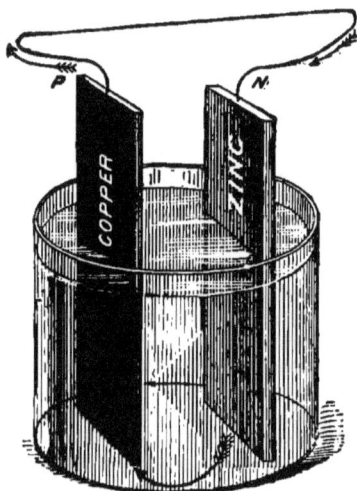

Fig. 1.

the copper or insoluble plate, and, on account of the direction
of the flow within the solution, the copper plate is called electro-negative, and the zinc electro-positive.   But the wire
attached to the electro-positive plate is called the negative
pole, and in all batteries in practical use this is a zinc plate
and generally called the negative plate synonymously with the
negative pole of the battery.   The reason for this will be here-
after explained.

The force in virtue of which this current flow is produced is
called the electro-motive force or difference of potential.
These terms are used synonymously, but it is customary to
apply the term "difference of potential" to a particular case

of electro-motive force. The arrangement of two metals, as shown in Fig. 1, is called a simple voltaic element or couple, or commonly a single cell, and several of these cells constitute what is known as a galvanic battery. The current produced by this battery has various names, such as the battery current, the constant current, voltaism, and galvanism, but is generally known as the galvanic current. The conditions under which a current of electricity is created in the above cell may be illustrated by reference to the conditions which exist and cause a flow of water between two reservoirs when one is on a higher level than the other. If they are connected by a pipe, the water will flow from the higher reservoir to the lower until the level is the same in each. If, however, the lower reservoir is so large as not to have its level affected by this flow from the upper, and we have a means of keeping the upper reservoir full, then we shall have a continuous flow from the upper to the lower reservoir. Again, if we charge an insulated body, say a metal sphere, with positive electricity and charge a similar sphere with negative electricity, and then connect the two with a piece of wire, a current will flow from the positively charged sphere to the negatively charged sphere, or from the higher to the lower potential.

As we speak of electricity flowing in the same sense as of sound or light traveling, we may adhere to the chemical view and state that the more soluble of any two metals, or the one most acted upon by the solution, is the electro-positive element, and that it is at a higher potential than the one less acted upon, and, as we have seen, the current flow is from the most corroded higher potential zinc plate through the solution of the cell to the less corroded lower potential copper plate. Bearing in mind the fact that the current always flows from the positive to the negative, we trace it from its starting point at the surface of the zinc plate through the solution to the copper plate which conducts it up to the connecting wire, which in turn conducts it back to the zinc plate; therefore, in the solution the zinc is positive, as the current runs from it, and the copper is negative, as the current runs to it, while outside the cell the copper is positive, as the current runs from it, and the zinc is negative, as the current runs to it. This explains why the positive pole of the battery is attached to the negative element or plate. We may state briefly that the electro-motive force is a measure of the difference of potential between the two elements of a battery cell. If we join several elements together, connecting the positive pole of one cell or couple to the negative of the next, we have what is called a compound circuit, or, speaking more generally, we have the cells connected for tension or in series

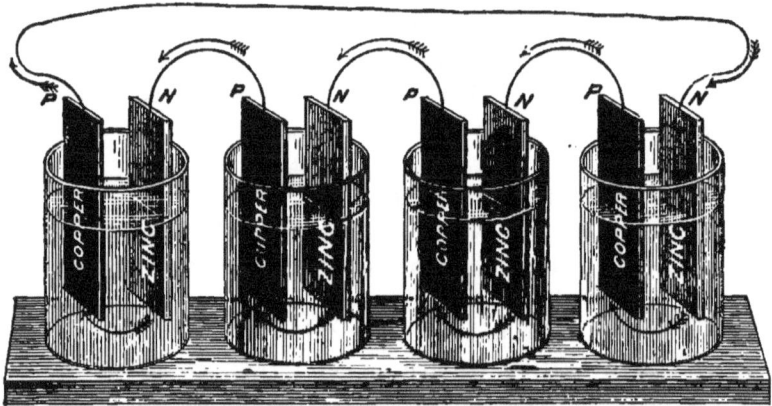

Fig. 2.

(see Fig. 2). If we have the same number of cells connected so that all the positive plates of the couples are connected together and all the negative plates together, the cells are said to be put up for surface, quantity, or parallel. In the first instance, if the wire connecting the positive and the negative elements be of large size, so as to offer practically no resistance to the current flow, the cells will cause no more current to flow through the wire than if one cell alone were used. In the second case, the (+) positive plates of all the cells being connected together and all the (—) negative plates connected together, the current flow will be four times as great as in the series tension or compound arrangement. In order to fully explain why this is so, and also to explain that with a wire or other body or conductor of electricity offering high resistance to the current flow the reverse of the above facts is true, makes it necessary to consider the laws which govern the phenomena of the electric current, and which as closely apply to the use of the electric current in the treatment of disease as in the various arts of electro-metallurgy, electric-lighting, and the furnishing of power for locomotion and other industries. The law which governs the production of the various results attained by use of the electric current is called Ohm's law, and is the corner-stone of all the progress in electrical arts and industries in the present century. If now we let E stand for electro-motive force, which will be remembered to be the initial force of the current, and R for resistance, which is the opposition offered to the passage of the current, and which may be divided into internal and external; the internal being whatever intervenes between the plates inside the cells, such as the liquid or porous cup, if one be used,—the external resistance being whatever intervenes between the plates outside the cells; being the conducting wire, and in the medicinal use of electricity that part of the body in the circuit; and C for current strength, or the rate of current flow that actually passes through the conductor. Ohm's law is then expressed by the formula, $C = \frac{E}{R}$, viz. $\frac{\text{The electro-motive force}}{\text{The resistance.}} =$ the rate of the current flow. It will be seen that, if two of these factors are known, the other may be readily obtained, for if $\frac{E}{R} = C$, and we have E and R known, we readily obtain C.

If C and R are known, then $R \times C = E$, and, finally, if C and E are known, $\frac{E}{C} = R$, letting $E = 4$, $R = 2$, then $\frac{4}{2} = C = 2$. Second, $C = 2$ and $R = 2$, then $2 \times 2 = E = 4$, and third, $E = 4$ and $C = 2$, then $4 \div 2 = R = 2$.

Polarization.—If the wire connecting the positive and the negative pole in the figure be one offering little resistance to the passage of the current, after a short time the strength of the current will greatly diminish, and finally cease altogether, or nearly so. If now we examine the copper plate carefully we shall find it covered with minute bubbles of hydrogen, which form an intervening layer between it and the solution of dilute acid in the cell; so we have a plate of hydrogen, practically, instead of a plate of copper, and the current is weakened.

The hydrogen collected on the copper plate also tends to react upon the sulphate of zinc formed by the dissolving of the zinc in the solution and to cause a deposit of the same upon the copper, thus still further diminishing the electro-motive force of the couple—instead of having a plate of copper and another of zinc, we in reality have two zinc plates. The weakening of the current from all of the above causes is termed polarization. To restore the cell to its original power, the solution may be agitated by moving the elements in the fluid, thus washing off the bubbles of hydrogen, or by blowing air in between the plates. Cautery batteries, which polarize rapidly, are so constructed that the plates can be moved around in the fluid or provided with a bulb for blowing air in them. Breaking the circuit and allowing the battery to rest for a moment will also cause the hydrogen to escape.

Local Currents.—If the zinc is not chemically pure, the impurities set up local currents, caused by the particles that are electro-negative to the zinc, forming minute closed circuits with it and destroying or diminishing the effect of the plate as a whole with the copper plate. This local chemical action caused by the impurities may be overcome by thoroughly amalgamating the zinc plate by immersing it in a vessel containing metallic mercury and dilute sulphuric acid, and then rubbing it with a swab, which gives a fine, even coating of mercury to the zinc, and it thus acquires all the properties of chemically pure zinc. The form of cell we have been considering was made by Volta in 1800, and also a series of similar cells called the " crown of cups "—being the first galvanic battery, or perhaps more properly voltaic battery, and, as essentially arranged by Volta, is in practical use to-day. Fig. 2 shows the crown of cups, the zinc and copper being soldered together, and the zinc of one pair dipping into the cell with the copper of the next pair, and so on, the terminal plates being single, one of copper and the other of zinc, as shown.

The principal improvements on this form of cells are chiefly in the substitution of different fluids, and the substitution generally of carbon for copper, except in cases where the salts of copper (chiefly the sulphate) are used, and here the two original elements are still retained.

The fluid generally used with the carbon is still dilute acid ; but with the addition of bichromate of potash salts, which tends to lessen the polarization, and at the same time the electro-motive force of the cell is greater, and the internal resistance less. The cell we have been considering belongs to a class we may call single fluid cells with no depolarizer, and is at the present day replaced by the zinc-carbon cell last named, more commonly called the bichromate cell or Grenet cell, and which belongs to a form of battery known as single fluid batteries, with liquid depolarizers. As it may be instructive to consider a few of the various forms of batteries, we will speak of one cell belonging to the same class as Volta's original zinc-copper-dilute-acid cell. This cell is that of Smee, devised forty years after Volta's crown of cups, and in which the copper is replaced by a plate of platinum, or generally by a platinized silver plate. The reason for platinizing the

Fig. 3.

silver plate is to cover it with minute points from which the hydrogen bubbles will free themselves and pass off, leaving the plate clean.

The cell is shown in Fig. 3, with a platinized silver plate, and two zinc plates, and the solution dilute sulphuric acid. This cell is not much used at the present time.

The next class of cells are those with the liquid depolarizer, which we may pass by ; simply referring to Fig. 4, which shows the ordinary construction of the bichromate cell as used for experiments, running faradic coils, and other purposes, where a strong current is needed for a short time only. The zinc is arranged to be lowered into the fluid when it is desired to use the current, and raised out when not in use. The solution is dilute sulphuric acid (one part acid to ten or twelve of

Fig. 4.

water) and three ounces of powdered bichromate of potash to each pint of the above solution.

The next class of cells is that of single fluid cells having a solid depolarizer. Fig. 7 represents the Barrett chloride of silver battery, the elements of which are a rod of zinc and a

rod of silver, the rod of silver being surrounded by a cylinder of fused chloride of silver. This cell is constant, but has
great internal resistance, and can not give a large current flow.

Fig. 5 shows the disque cell of Leclanché; with a zinc
rod, and the carbon, in a porous cell surrounded by granulated
manganese and small fragments of carbon. The fluid is a
saturated solution of chloride of ammonium. This is one of

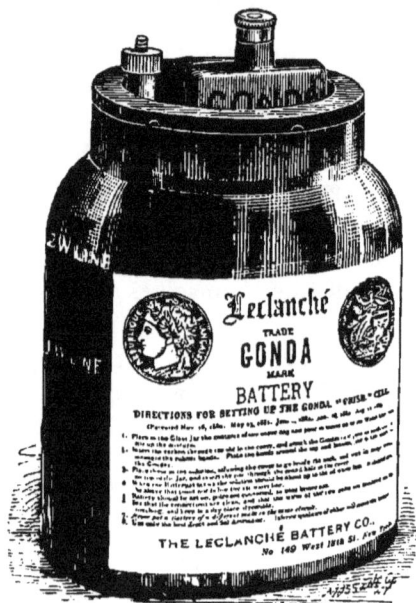

Fig. 5.

the most useful forms of battery known for intermittent work
of any kind, or for any purpose not requiring a large rate of
current flow long continued.

We now come to the third class of cells: those having a
liquid depolarizer surrounding the insoluble electrode only,
and separated from the zinc and the fluid surrounding it by a
porous diaphragm (generally an unglazed vessel of clay) commonly called a porous pot. In some of these cells the zinc is
placed inside of the porous pot, and the insoluble electrode
in the containing vessel with the depolarizing fluid, and in
others the zinc, generally in the form of a cylinder, surround-

ing the porous cell and the insoluble electrode in the porous cell, as first mentioned.

One of the most important and among the first two fluid batteries is that of Daniell. This cell, called "Daniell's cell," is the most constant cell known. It consists of a glass jar, a cylinder of zinc inside a circular porous pot, and a copper cylinder surrounding the porous pot. The fluid surrounding the zinc, which is thoroughly amalgamated, is dilute sulphuric acid. Around the copper is a saturated solution of copper sulphate. The action of this cell is as follows : when the circuit is closed the hydrogen set free at the copper plate meets the sulphate of copper, which is reduced to metallic copper and deposited on the copper plate. The changes in the chemicals in the porous cell and containing vessel are almost without number.

We shall find it necessary to mention but a few of the two-fluid cells, as we insert a table on page 22 showing the different fluids used from time to time, and their influence on the electro-motive force of the various cells. The Daniell cell has a low electro-motive force and a high internal resistance.

In 1839, Grove invented the cell called after his name—the Grove cell. In this cell the porous pot contains strong nitric acid with a plate of platinum, and, in the containing vessel, zinc in dilute sulphuric acid. This cell is one of the most powerful known, has a high electro-motive force, and low inter_ nal resistance. It, however, gives off strong fumes of nitric acid. The objection just mentioned (the giving off of the poisonous fumes of nitrous oxide), can be and generally is avoided by the use of a different depolarizing liquid. This modification is called the carbon or bichromate of potash battery. The zinc, in the form of a cylinder, surrounds the porous pot, a plate of carbon being placed in the porous pot and surrounded by a mixture of sulphuric acid 100 parts, water 25 parts, and pulverized bichromate of potash 12 parts.

For medical use we should choose a battery having a high electro-motive force, and a low internal resistance.

The following table gives the electro-motive force and the elements and fluids used in all of the various forms of batteries in use to-day :

ELECTRO-MOTIVE FORCES OF VARIOUS CELLS.

	+ Plate.	Porous Cell.		— Plate.	B. A. Volts.
Daniell,	Zinc amalg.	Sulphuric Acid, 7½ to 1	Saturated solution of copper sulphate	Copper	1.079
"	"	22 to 1	"	"	0.978
"	"	"	Nitrate of copper saturated	"	1.000
"	"	"	Sulphate of copper	"	0.909
" P. O. Standard,	"	Sulphate of zinc saturated solution	Saturated solution	.. " ..	1.079
Grove,	"	Sulphuric acid, 7½ to 1	Nitric acid (fuming)	Platinum	1.956
"	"	Salt water	Nitric acid, sp. gr. 1.33	"	1.904
"	"	Sulphuric acid, 22 to 1	"	"	1.810
"	"	Sulphate of zinc	"	"	1.672
Bunsen,	"	Dilute sulphuric acid	Nitric acid	Carbon	1.734
Smee,	Zinc	Sulphuric 1, water 7		Platinised sil.	.48..1.1
Walker,	"	"		" carbon	.5..1
Callan,	Zinc amalg.	Dilute sulphuric acid	Nitric acid	Cast iron	1.700
Poggendorf,	"	"	{ Bichromate of potash	Carbon	{ 1.796 / 2.028
Marié Davy,	"	Sulphuric acid, 22 to 1	Paste of sulphate of mercury	"	1.524
"	"	Dilute sulphuric acid	"	"	1.33
Leclanché,	"	Solution of sal ammoniac	Binoxide of manganese	"	1 48
De la Rue	Zinc	Chloride of ammonium		Silver, + AgCl	1.030
Skrivanov (pocket form)*	"	Solution 75 caustic potash to 100 water		" "	1.4 to 1.5
Becquerel,	Zinc amalg.	Sulphate of zinc	Sulphate of lead		
Niaudet,	"	Common salt	Chloride of lime	Lead	0.55
Duchemin,	"	"	Perchloride of iron	Carbon	1.65
" Secy.battery	Platinum	Dilute sulphuric acid	Dilute sulphuric acid	Lead	1.541
				Platinum	1.79
Latimer Clark, (standard cell)	Zinc amalg.	Sulphate of zinc	Paste of sulphate of mercury	Mercury	1.45
Howell's manganese; internal res.= 1 ohm (Hockin),	Zinc amalg.	Ammonic sulphate, 25 grammes crystallised salt to 1 litre water	Sulphuric acid, 1 acid to 5 parts water	Carbon + manganese dioxide + manganese sulphate	2.04
Higgin's cascade, internal res.=.170 ohm,	Zinc in mercury	Chromic acid	Sulphuric acid	Carbon	1.9
Thame's,	Zinc	Dilute sul. acid	Nitric acid + CrO2 Cl2	Carbon	2
Bennet's, internal res.=5ohms	"	Potassium hydroxide (KHO), with distilled water	Damped with (KHO) and water	Iron can with iron borings	1.3
Lalande-Chaperon,	Zinc amalg.	Caustic soda solution	Oxide of copper or " copper scale "	Iron	1
Faure's secondary battery,	Lead plate coated with minium	Dilute sulphuric acid	Dilute sulphuric acid	Lead plate coated with minium	2. to 2.2
Sellon-Volckmar	Lead plate primed with minium	Sol. sulphuric acid, sp. gr. 1100°	Sol. sulphuric acid, sp. gr. 1100°	Lead plate primed with minium	2.25
Planté,	Lead	Dilute sul. acid	Dilute sul. acid	Lead (spongy)	2 to 2.2

Note—These E.M.F.'s are 1.1 per cent. too high, and should be multiplied by .9889, the ratio of the B.A. unit to the legal ohm.

Special Forms of Galvanic Batteries.—Fig. 6 shows a very convenient form of battery manufactured by Messrs. Waite & Bartlett, of this city, which I use for all work when it is necessary to carry a battery about, as it is simple in construction

* Used for electric " star " lights.

and easily managed, and not liable to get out of order. It is shown in the figure with its front, which is in form of a door, thrown open, exposing the zincs, carbons, and jars to view, so that the condition of all may be easily and thoroughly inspected at any time. The tray or drawer containing the cells has a flush ring in the front, by means of which the drawer may be pulled out and the cells filled or emptied, or the condition of the solution examined. The solution used is sulphuric acid 1 part, water 16 parts, bichromate of potash 2 parts, and 15 grains of bisulphate of mercury. If the fluid

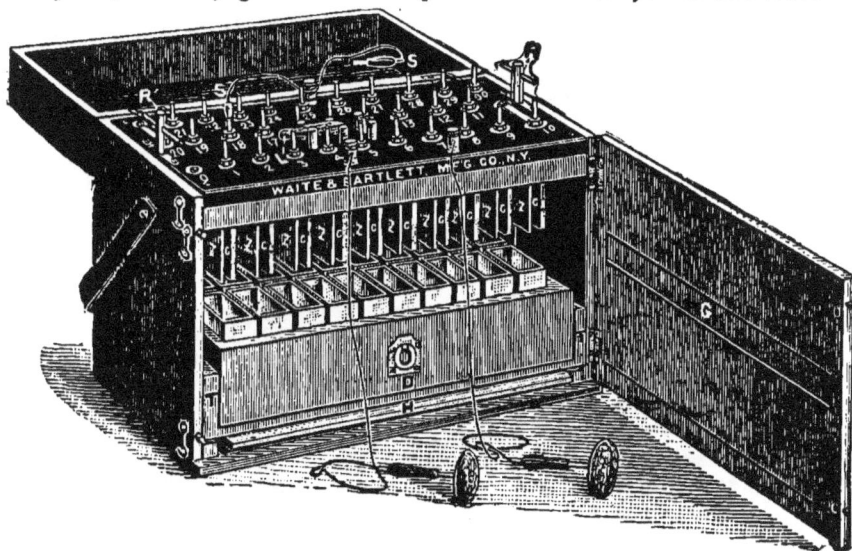

Fig. 6.

retains any yellow color it is still in condition for use ; but if green or black must be removed. At the bottom of the box is shown a rubber-padded board, called a hydrostat, which may be placed padded side down on the top of the cells. In the back of the box is a groove similar to that shown in the door at *g* and at the same height ; after placing the board on the top of the cells, the door is closed and the two lifting-rods are screwed up, thus raising the cell-tops against the padded board and sealing the cells, as the board can not be pressed above the grooves. To put the battery in operation, the two lifting rods (RR) are raised and then given a quarter turn. This operation raises the platform and tray of cells so

that the zincs and the carbons are immersed in the solution
of the cells. The selecting of the cells will also be found convenient and easy of manipulation. At the back of the top
board is attached a bifurcated cord, to the free ends of which
are attached two sockets. To draw on the cells it is only
necessary to place one of the sockets on the pin marked "one,"
when the current from one cell will pass to the binding screw
(SS'), and thence to the electrodes. If more than one cell is
desired, place the other of the two sockets on the post marked
"two" and then remove the socket from pin "one," and so on.
The battery is also provided with a commutator or polarity

Fig. 7.

change by means of which the positive pole may be made to be
at either of the two binding-posts. I have described this
instrument at some length, as I use it personally outside of my
office work.

Fig. 7 represents the Barrett chloride of silver battery,
sold by Messrs. G. Tiemann & Co., of this city. This is a very
convenient form of battery to carry around, and, when carefully used, answers the purpose remarkably well for light work,
but is not suited for the department of gynæcology, electrolysis or heavy work.

In my office work I use the form of switch-board, current selector, etc., as shown in Fig. 8.   This switch-board is design-ed for economy of space in office and for convenience as well. The current selector is universal, and by it any number of cells from any part of the series may be brought into the circuit.

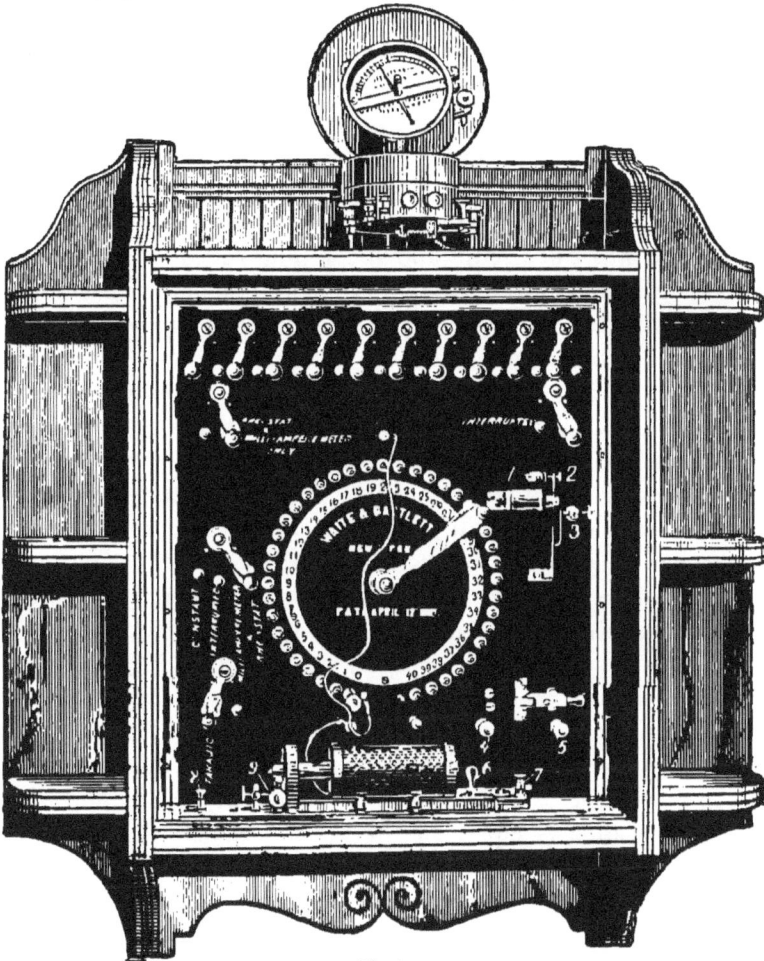

Fig. 8.

The selecting lever slides over pins, instead of the buttons used in most switch batteries.   The socket, (shown on pin in the figure) is removable, and the lever always being kept in advance of the socket, the pole changer (*P*) always points toward the positive pole ; but, should the socket be placed in

advance of the lever, it points toward the negative pole. The
number of cells in the circuit is determined by the difference
between the number on which the socket and lever rest. The
board is provided with a commutator, as described on
the portable form; an automatic rheotome is also provided, by
means of which an intermittent or pulsating current is produced. This is of great advantage in the treatment of constipation or other conditions where we wish to get the deep
penetrating and stimulating effects of the galvanic current.
A means of modifying the current independently of the current
selector is provided in the water rheostat. By means of the
current selector a number of cells from any part of the series
is brought into the circuit and the sliding rod of the rheostat
raised. No perceptible current is felt on placing the sponges
on the patient. By gradually pushing down the rod of the
rheostat the current may be more gradually increased than by
the adding of one cell at a time. This will be found of great
service in treating certain forms of spasms, neuralgia, tinnitus,
etc. In addition to the water rheostat, I have a wire rheostat,
consisting of coils of German silver wire accurately measured,
and by means of which, in conjunction with the milliampere
meter, we are enabled to measure the resistance of the patient.
There are two methods by which this may be accomplished.
First, by turning on cells enough to give a deflection whose
value is, say, twenty milliamperes, we now (leaving the number of cells unchanged and the electrodes *in situ* on the
patient) throw in the circuit by means of the rheostat sufficient resistance to reduce the deflection on the galvanometer
to ten milliamperes, or one-half as much as before, and this
added resistance is equal to the resistance of the patient.
The reason this is so will be apparent when we consider that
with a given electro-motive force the current flow in any circuit is inversely in proportion to the resistance; or, in other
words, the greater the resistance, the less the current. As
we have just seen, the resistance of the patient causes a certain
current; we double this by adding enough resistance to reduce
the current one-half, hence we must have added an amount
just equal to the resistance of the patient. Second, another
way, and by the use of which it does not become necessary to

keep the current on the patient only during the regular treatment, is by substitution, and for this method of measurement
the switch-board has special devices. After getting the deflection of twenty milliamperes, as in the first case, by any suitable
number of cells, by simply turning a switch, the patient is cut
out of the circuit and the rheostat substituted for the patient.
It now simply becomes necessary to throw into the circuit sufficient resistance to produce the same deflection as was produced by the resistance of the patient, when the resistance
must equal that of the patient. It will be observed that the
same law of the ratio of resistance to current with a constant
electro-motive force holds true here, and the substituted
resistance, which is a known resistance, equals that of the
patient, because the current is the same in both cases. It
must be remembered that the number of cells in use during
the treatment of the patient and the observation with the
rheostat must be the same. This key-board also contains a
faradic battery with the coils of the Du Bois-Reymond pattern
and of the Engleman dimensions, and will be described further
on. Fig. 9 represents a cabinet battery with the same key-
board.

Milliampere Meter.—It becomes necessary to speak of the
accessories wanted in connection with the use of a galvanic
current, and we must place the milliampere meter at the head
of the list, as it is as important as the graduate and balance
of the apothecary, and, indeed, we may say no accurate or
repeated dosage or treatment can be given electrically without
it. A milliampere meter is a galvanometer, whose deflections
have a definite value, and may be called a graded or absolute
galvanometer. We may now refer to our original figure
again, and remember that on the wire being placed in the ends
of the zinc and copper plates, we observed that it became
warm and that it deflected the needle suspended over it to
the right or east. If the wire, instead of passing under the
needle, passes over it, the deflection will be greater, and if
the wire be made into a coil, the deflection will be still
greater. Such a coil with a needle suspended within it is
called a multiplier, on account of the added or accumulative
effect of each additional coil or spiral of wire. By placing a

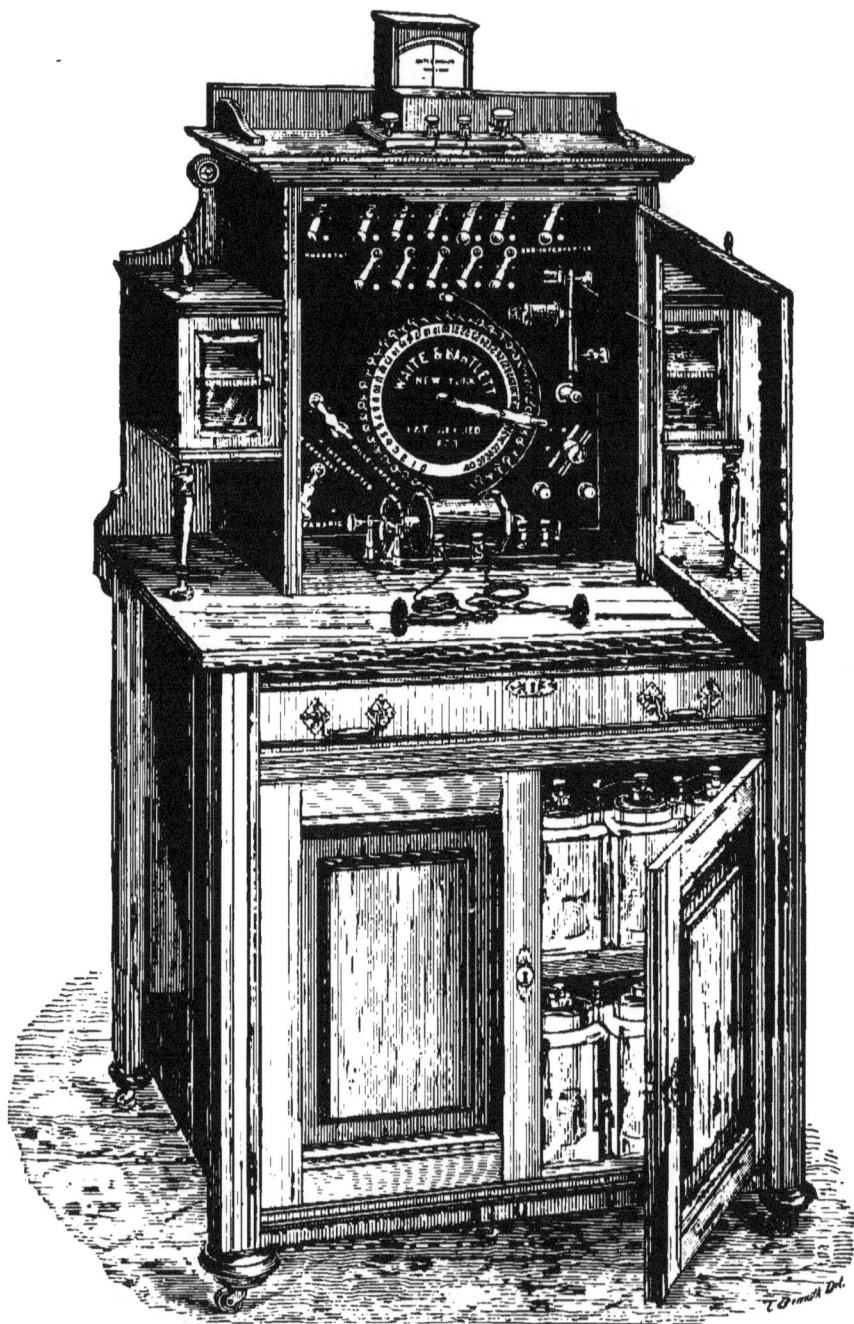

Fig. 9.

cardboard under the needle on which to engrave or mark a scale, we have a galvanometer, which, by proper calibrating, may become a milliampere meter. In some forms of galvanometers the wire simply passes around the needle once, and generally in this form of instrument the coil is circular. If the coil has such a diameter that a magnetic needle hung in its center, and the said coils placed so that its winds or turns of wire are parallel with the direction of the needle, the currents sent through the coil cause the needle to deflect in proportion to the tangents of the various angles of the deflections, provided the diameter of the coil and the length of the needle are proportioned to each other. This instrument is then called a tangent or proportional galvanometer. This may

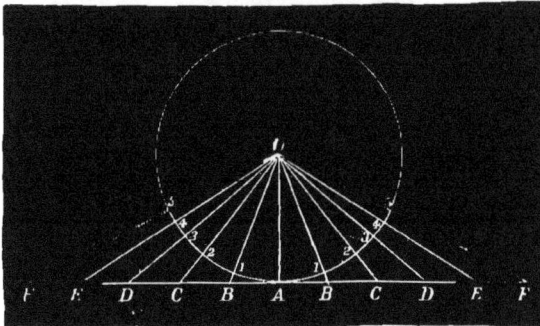

Fig. 10.

be explained by the following diagram (Fig. 10): Let the length (*AB*, *BC*, *CD*, etc.) along the line *AF*, which is tangent to the circle at the point *A*, be made equal to each other. Draw the lines *oA*, *oB*, *oC*, cutting the circumference of the circle at the points 1, 2, 3, 4, 5. Then the numbers 1, 2, 3, etc., are proportional to the tangents of the angles *ao1*, *ao2*, *ao3*, etc. It can be plainly seen by the above diagram that the spaces between the lines drawn from the center of the circle (*o*) to the various points (*A*, *B*, *C*, *D*, etc.,) on the line *AF*, are equi-distant, but at the points where they cut the circle, as 1, 2, 3, etc., they rapidly approach each other as they are drawn further from the vertical line *oa*. In a tangent galvanometer the diameter of the coil is very large in proportion to the length of the needle.

It has been found that for a needle about one inch or less in
length the diameter of the coil should be about twelve inches,
or about twelve times the length of the needle, to deflect it
in proportion to the strength of current sent through the coil.
Suppose we find that a current strength will deflect the
needle from its resting-point (A) to point 1 on the circumference of the circle, then a current twice as strong should deflect
it to the point 2, since at that point we find the line o C to
cut the circle, C being the same distance from B that B is
from A. Three times the current strength would deflect the
needle to 3, and four times the strength to 4, etc.   If

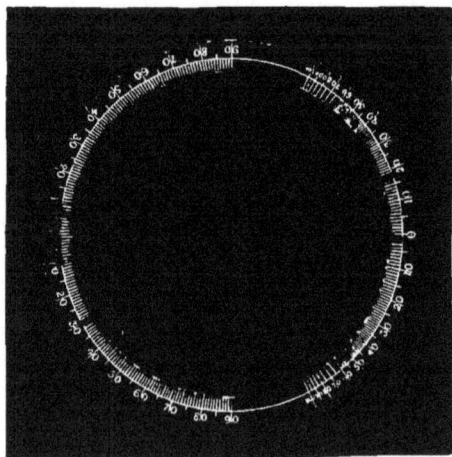

Fig. 11.

we should find, however, that five times the current strength
did not cause a deflection on the scale to point 5, but that
it rests about midway between the points 4 and 5, we
should say that we had reached the limit of our instruments
proportionally, or that our instrument was proportional to the
tangents of the angular deflection to 40°, if the point 4 be
40° from A. Fig. 11 shows a dial divided into degrees on the
left side and tangents on the right.   If we let the first sixty divisions on the tangent scale represent sixty milliamperes, the
lines come very close together, and, to avoid confusion, we
sub-divide the next ten at five, for in the use of such a strong
current as sixty milliamperes, we do not need to be so particu-

lar in reading a single milliampere. If we had access to a standard galvanometer we might, by comparison, construct a milliampere meter by the grading of a scale on any galvanometer, having the standard galvanometer and the one of unknown deflective value in the same circuit with any suitable battery, and a means of varying the current to produce deflections so that the needle of the standard instrument will rest over the divisions of its scale, and marking the various positions of the needle of the new instrument at the same time. These divisions on the new instrument have the same value as those of the standard instrument.

One very annoying thing is the oscillation of the needle after changing the current strength. To avoid this, many devices have been used, such as air-vanes and paddles dipping in water or alcohol. Solid masses of copper, which, as well as several other metals, have the power of checking the oscillations on account of the magnetic currents set up by the movements of magnets in close proximity thereto, are also used. The instrument represented in Fig. 12 is the one used by the author. The deflections are in proportion to the current strength, as far as the divisions are marked on the scale. The divisions represent primarily only five milliamperes, and are far apart and subdivided to allow of delicate currents being used and measured accurately. The value of the readings of the scale may be increased ten or one hundred times by means of thumb-screws provided for that purpose, so that the readings may mean 1, 2, 3, or 10, 20, 30, or 100, 200, 300, etc. These thumb-screws are so arranged that when the one marked "ten" is screwed down it throws into the circuit a conductor having only one-tenth of the resistance that the coils which influence the deflections of the needle have, therefore it conducts off nine-tenths of the current, and only one-tenth of the current goes through the coils which influence the needle, consequently the needle only deflects one-tenth of the current strength. The screw marked "one hundred" conducts off ninety-nine one-hundredths of the current strength, and in the same way the needle deflects only one one-hundredth of the current. By this means the instrument is capable of measuring currents varying from one-fifth of one to five hundred milliamperes. The needle of this instrument consists of a strong horse-shoe magnet in the shape of a bell and called the bell magnet ; it is much stronger than the ordinary compass-needle, and consists of a piece of steel rod drilled nearly through its length and then sawed or slotted diamet-

Fig. 12.

rically nearly as far as the drilling. This is carefully tempered, hardened, strongly magnetized, and then hung in the center of a copper cylinder which is bored to the depth of the magnetic length, the opening being slightly larger than the magnet. To the top of the magnet is secured the aluminum pointer or index that swings over the dial. Outside of the copper block on either side are two coils of wire. These coils

are so placed that the plane of their windings is parallel with
the magnetic meridian of the earth and with the polar
length of the needle. In using this instrument it must be
placed perfectly level, so that the side of the magnet does
not touch the copper, and so that the north pole of the pointer,
when at rest, points directly to o.

Faradic Battery.—We have seen that one of the properties
of the wire which connected the copper and zinc plate of a
battery cell is induction, and it is this inductive property
that gives us the battery known as the faradic battery, which
is simply an electro-magnetic machine. If we wind a coil of
wire as thread is wound on a spool, the wire being covered
with silk so that one turn or spiral may not be in contact with
another, and then wind another coil in the same manner, having it large enough so that it will slide over the first coil, and
attach the two ends of the first coil to the plates of a battery
cell and the two ends of the second coil to a galvanometer,
we find that when the second coil is rapidly shoved over the
first coil the galvanometer needle is deflected, but it immediately returns to zero, where it remains until the second coil
is suddenly removed, when the needle is again deflected but in
the opposite direction. This first coil is known as the primary coil and the second or outside coil as the secondary. We
may further intensify the deflection of the galvanometer
needle by placing inside the primary coil an iron bar or a
bundle of small iron wires, completely filling the opening.
The action of the iron bar is first to concentrate the lines of
force set up by the coils of wire as a lens concentrates rays
of light, and currents are also set in action or induced in the
iron core itself, which on being rapidly demagnetized by the
breaking of the connections with the battery greatly increases
the strength of the induced or extra current momentarily set
up in the coils. If, after filling the primary coil with the
iron wires, we take in our hands the ends of the wire of the
outside coil and then attach the ends of the primary coil to
our battery cell, we shall perceive little or no sensation, but if,
having once closed the circuit as aforesaid, we suddenly break
the connections of the primary coil with the voltaic cell, we
will receive a decided and disagreeable shock, and if the coils

are of sufficient size contractions of the muscles in the arms
and hands will be produced. If we make and sever this connection very rapidly we produce a continuous contraction of
the muscles of the hand and wrist. In batteries in practical
use this is done by an automatic interrupter called the
rheotome.

The connection of the primary coil with the cell is by way
of the post at the left in Fig. 13, thence to the flexible spring,
and from the spring to one end of the primary wire, the remaining end of which is attached to the other element or plate
in the voltaic cell. If the cell is filled with fluid and in proper
order, a current flows along the primary wire in the course just
mentioned, and, as it winds around the iron core, that becomes
strongly magnetized, and the spring, having a piece of iron

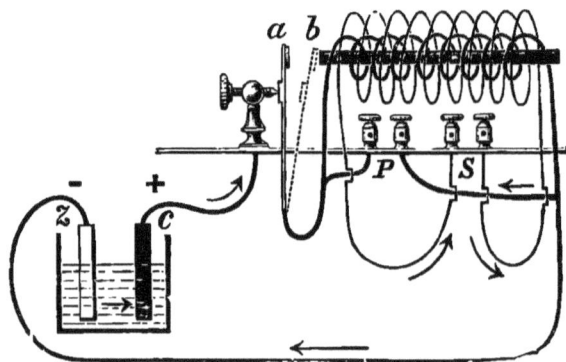

Fig. 13.

affixed to its upper extremity, is attracted toward the end of the
iron core or primary coil at *b*, until the spring breaks its connection with the end of the screw, thus breaking the circuit and
the current ceases to flow from the cell. At the same instant
the magnetism in the core ceases, and the spring by its own
elasticity returns to its original position in contact with the
point of the screw, when the circuit being again made the spring
is again attracted, when it is again broken and flies back again,
and so on. This is repeated very rapidly, and this rapid
breaking causes the extra current to give these shocks, and
when in rapid succession constitutes the faradic current.
Shocks can be also obtained from the primary coil, but they
are much weaker than those from the secondary. The

strength of these shocks may be varied in five different ways: First, the shocks are decreased by removing the iron core and strengthened by replacing it. Second, by placing a brass or copper tube over the iron coil. Third, by placing the same tube between the primary and secondary coil. Fourth, by placing the same tube over the secondary coil. In

Fig. 14.

the last three cases the shocks are strengthened as the tube is withdrawn and lessened as it is replaced over the coils. If this tube were divided its entire length so that it is not a complete circle, it will have no effect on the strength of the shocks. In all of the above cases the primary and secondary currents are affected in the same manner. Fifth, the shocks are modified by the distance the secondary coil is shoved over the primary.

The shocks from the secondary coil is much greater when it is
shoved over the entire length of the primary, and will be gradually lessened as the secondary coil is withdrawn. If the two
ends of the secondary coil are not connected together it has
no influence whatever on the primary coil, as it acts the same
as the split tube ; but when they are connected, the shocks
of the primary coil are greater as the secondary coil is withdrawn and less as it is replaced.

On the last principle the Du Bois-Reymond coil is made.
Fig. 14 represents the Engleman battery, which is constructed with three secondary coils made of wire of different
lengths and diameters, and will be seen further on to possess
special qualities. The secondary coil is provided with a switch
which disconnects the two ends of the secondary coil when it
is desired to use that coil, and connects them when it is desired to use the primary coil. This battery possesses one great
advantage over any other : every battery is the same, thus
giving us as near a uniform faradic battery as it is possible to
have, and if it was used by all physicians, some idea could be
obtained regarding the strength of current used in the reported
cases. It also has a slow and rapid vibrator, which is quite
essential in the treatment of various diseases.

Electrodes.—Only those electrodes which are in common use
will be described in this place, the special electrodes being
described with the various treatments and operations in which
they are employed.

The first is the universal handle (Fig. 15). This is simply a
brass rod running through a wooden handle with the split tip
attachment on one end for the rheophore or cord leading from
the battery and a threaded socket on the other. This socket
and thread is of a universal size, which gives the name to the
handle. If a handle is ordered from any of the leading electrical manufacturers (with two or three exceptions) in the
United States, this screw will be of the same size and will
consequently fit any instrument made for a handle.

The interrupting handle (Fig. 16) is the same as the universal
handle, except the brass rod running through the wood is
divided, and the circuit is completed by pressing a spring with
the thumb or finger of the operator. This handle is indis-

pensable for making an electrical diagnosis, and is very convenient in treating the motor points or whenever a sudden interruption of the current is desired.

Fig. 17 represents a metal electrode two inches in diameter, which is covered with sponge and fits the universal handle.

Fig. 15.

This electrode is used for a great number of applications and is spoken of in special therapeutics as the " ordinary electrode."

. Fig. 18 represents a flexible hand-electrode. These electrodes may be of different sizes and are the most useful electrode for all purposes in the market. It is composed of a sponge surface, and a rubber cloth back to protect the clothing from the dampness of the sponge. Between the sponge and the rubber cloth back is a plate of soft zinc or copper, to which is attached the rheophore.

Fig. 19 represents a sponge-covered foot-electrode, which is made in a similar manner to the flexible hand-electrode.

Fig. 20 represents the uncovered metal foot-electrode,

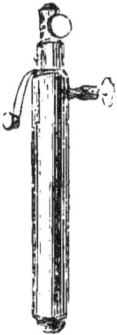

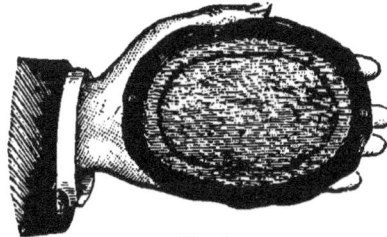

Fig. 16.   Fig. 17.   Fig. 18.

which is to be preferred in static electro-massage, and may be covered with chamois for other uses. If only one is to be had, the latter is to be preferred for all purposes.

Fig. 21 represents the spinal electrode, by which the spine can be treated along its entire length without wholly removing the clothing.

Figure 22 is taken from Erb, and the cut illustrates the full size of each electrode. The three to the left-hand of the page

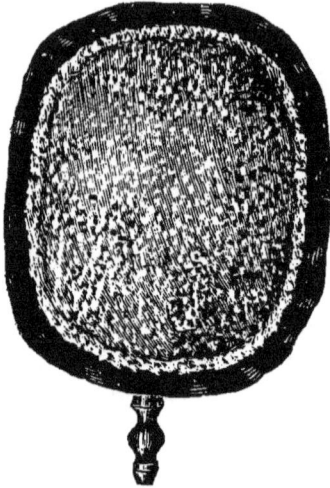

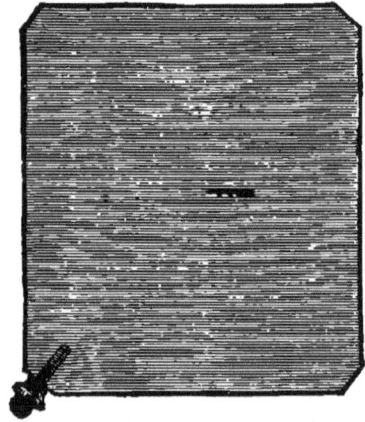

Fig. 19.   Fig. 20.

represent Erb's nerve electrodes covered with absorbent cotton. These instruments fit the universal handle, and are the best known to the author for reaching the motor points either for diagnostic or other purposes.

Fig. 23 represents Apostoli's clay electrode. This may be made by the physician. The best material to be obtained for making it is the finely ground clay used by artists for

Fig. 21.

modeling, which can be obtained at stores dealing in art materials. A piece of ordinary muslin may be used for the sack, but a towel which has been worn until it has become soft and smooth is to be preferred. This should be so cut that, when folded upon itself, it will be of the required shape and size. The edges are sewed together, leaving a space large enough for the hand to enter. The clay, well moistened, is packed carefully in the sack until it is from one to one and one-half inches in thickness. A brass plate sol-

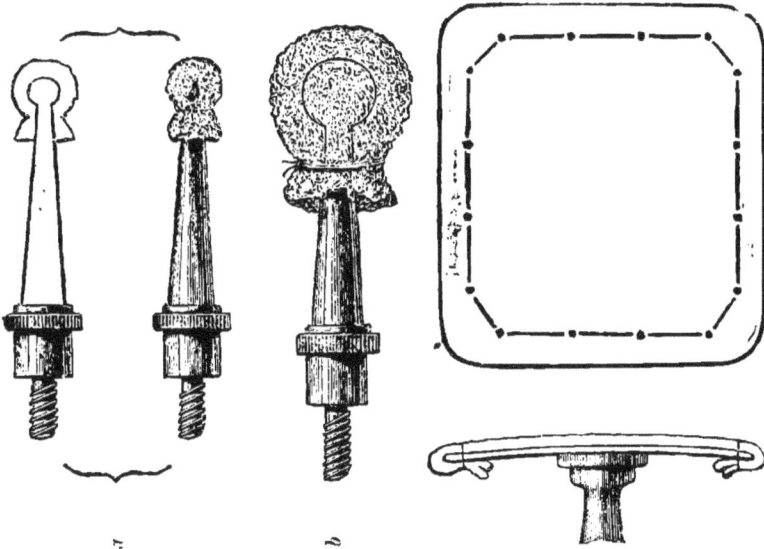

Fig. 22.

dered to one end of a copper wire, with a connector on the other end (as shown in cut), should now be imbedded in the clay and the opening in the sack closed around the wire. This electrode must be kept moist, which can be readily done

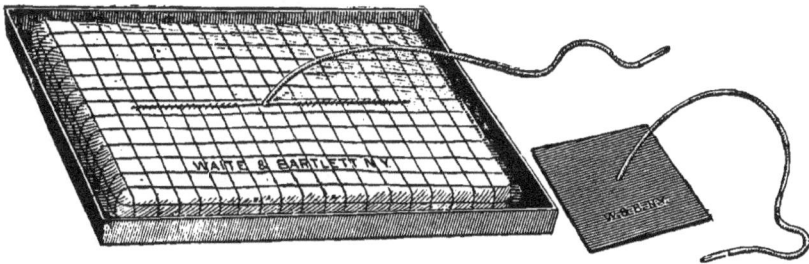

Fig. 23.

by keeping it in a little water in an ordinary dripping-pan. It will then always be ready for use, and is certainly the most effectual one the author has ever used, when very strong currents are given. In a recent conversation with Dr. A. D. Rockwell the author was informed that he placed a thin layer of absorbent cotton between the clay and the sack. This possesses many advantages. The absorbent cotton retains the moisture and thus keeps the clay in good condition; it is softer on its surface, and is more easily adapted to the abdomen; by using hot water the absorbent cotton becomes heated so that no unpleasant sensations are produced. To warm the ordinary clay electrode when it is used, it should be wrapped in a large wet towel and placed in an oven for a few minutes, when it will steam through and become thoroughly warmed without drying it.

A very cheap and convenient form of electrode is made of a wire mesh, which is cut the size desired. A split tip is soldered on one end and the edges dipped in hard solder, so as to prevent straggling ends of the wires from getting loose and pricking the patient. These are covered first with a layer of absorbent cotton and then with chamois. These electrodes can be bought in all sizes by the dozen from electrical manufacturers, and are the most convenient form of large electrodes.

Cautery Battery.—A cautery battery is so constructed that a large quantity of electricity is given off with small tension. Large plates are used, and are so arranged that they can be connected up for surface or quantity when it is desired to use a cautery knife in which the resistance is very slight, and consequently requires but little intensity, and in series when it is desired to use a large loop which has more resistance, and consequently requires more intensity.

Fig. 24 represents the Piffard cautery, which for all purposes is the best instrument ever used by the author. The figure represents it with its elements suspended above the fluid. This battery is so constructed that it can be rocked when in use, thus forcing the fluid through the openings in the outside zinc plates and striking the inside platinum plate, washing off the hydrogen from that plate and depolarizing the battery. The top of the battery is shown in Fig. 25. When we wish to connect the battery for quantity, we turn down the screws marked q. This connects the battery into two cells of three each, or, in other words, it connects the three cells of one side up for quantity—that is, zinc to zinc and platinum to platinum—and the three cells of the other side the same, and then the two sides are connected up in series. If we wish to connect the battery up for tension, or in series, we turn up the screw marked q and down the ones marked l. This connects the whole six cells in series. In using this battery one cord should be placed over the socket marked o, and, if it is connected for series, the other may be placed over the one marked 1, 2, 3, 4, 5, or 6, according to the number of cells desired. If the battery is connected for quantity, one of the cords should be placed over the post marked

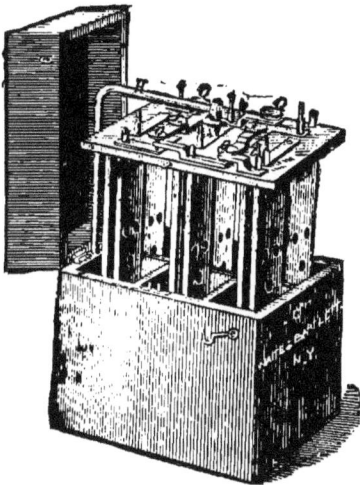

Fig. 24.

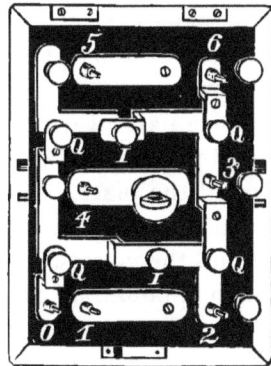

Fig. 25.

o, the same as before, and the other placed over the one marked 6.

The care and preparation of a cautery battery for an operation will be given in Chapter VII.

Static Electricity.—This form of electricity is sometimes called frictional electricity—the latter being one of the most common means of producing the same. It was known to the ancients that when amber was rubbed with silk it would attract light bodies, such as lint or small particles of thin paper.

In the year 1800, Volta, as well as inventing the first galvanic battery, also invented one of the most simple and inexpensive of electrical machines. This was an electrophorus, and from it emanated the various static machines, which are simply a continuous electrophorus. The electrophorus consists of a plate of resin ten or twelve inches in diameter and about one inch thick. This is generally on a metal surface, or may be put in a wooden mold lined with tin foil, which is called the form. We are not confined to a resinous plate, but may use a plate of glass or polished disk of hard rubber, and various other good insulators having a smooth surface. The other part of the instrument consists of a metal disk somewhat smaller in diameter than the resinous plate and provided with an insulated handle of glass or rubber. In order to work the apparatus it should be slightly warmed, to drive off all moisture. We then rub the surface of the resinous disk with a dry catskin, by means of which it becomes charged with negative electricity. If we now place the metal cover on the resinous plate the neutral electricity of the cover is decomposed into its positive and negative electricity, the positive part of which is attracted to the under surface of the metal plate and the negative electricity repelled to the upper part of the plate. If connection is now made with the upper surface of the plate the repelled negative electricity will pass off and the cover will remain charged with positive electricity, which is bound or attracted by the negative electricity produced on the upper surface of the resinous plate by the friction of the catskin. On the under side of the resinous plate is positive electricity repelled there by the negative electricity on the upper part of the resinous disk. As the resinous cake is a good insulator, the positive electricity of the cover does not unite with the negative electricity on its upper surface, and on the under side of the resinous cake the positive electricity induces or attracts negative electricity from the metal plate on which the cake rests, which latter, however, immediately passes to the earth. If we now raise the metal disk by the insulated handle, the positive electricity, being taken away from the attractive power of the negative electricity on the upper surface of the disk, distributes itself over the entire surface of the plate, and we may say the plate is positively charged. If a conductor be brought near the metallic plate, such as the knuckle, a spark will pass from the plate to the hand, and the metallic plate will return to its neutral state and exhibit no electrical phenomena. If we now place it again upon the resinous disk and then touch the upper surface with the finger, draw off the repelled negative electricity and then raise the plate again by the insulated handle, we obtain

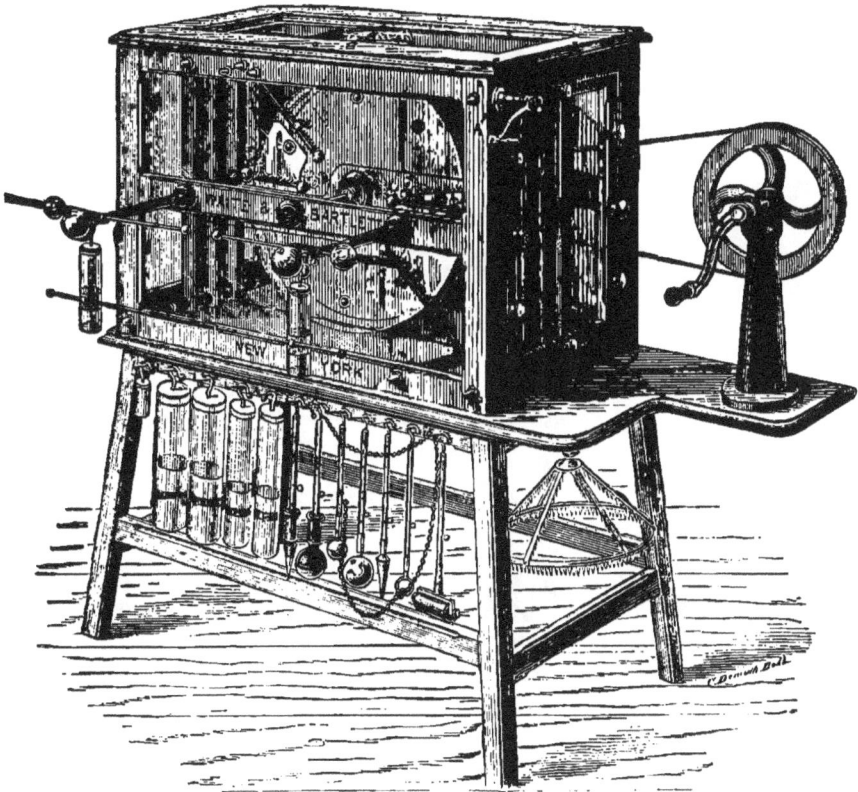

Fig. 26.

another positive charge from the plate, which may be drawn off by the hand or any other conductor, or which may be stored up in a Leyden jar by presenting the knob connected with its inside coating to the plate. The metallic plate on which the resinous cake rests makes the machine more active than it can be made without it. The reason of this appears to be that the positive electricity on the lower side of the resinous cake, being drawn away through the metal plate at

Wooden Ball.

Small Brass Ball.

Large Brass Ball.

Carbon Electrode.

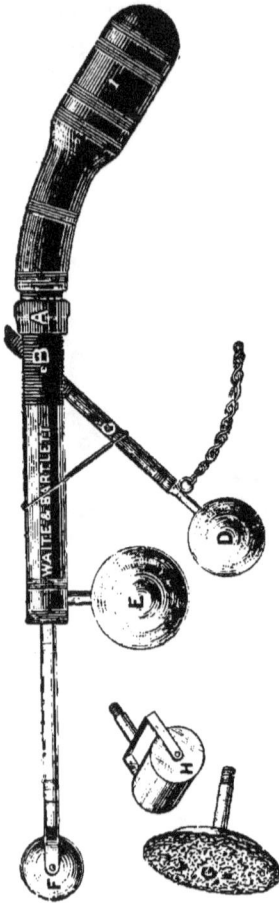

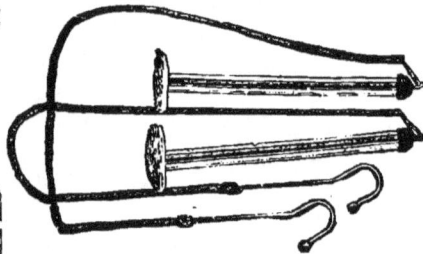

Handles and Sponges.

Massage Roller.

Morton's Pistol Electrode.

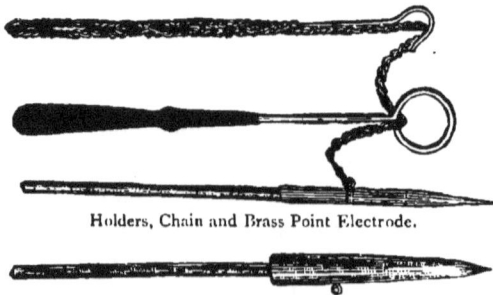

Holders, Chain and Brass Point Electrode.

Wooden Point.

The Spray.

Rubefacient.

Fig. 27.

the bottom, has a reactionary effect on the negative electricity on the upper surface of the resinous disk.

The plate electrical machine is but a development or more perfected form of the electrophorus.

The battery in use to-day is known as the Holtz machine, which was invented by Dr. Holtz of Berlin, in 1865. This machine does not generate electricity by friction, but by induction, as did the electrophorus. It consists of two glass disks, one revolving and the other stationary.

There are metallic conductors in the form of combs, which conduct the electricity to the two poles. The principal improvement on the Holtz machine is in adding more plates, thus giving off a larger quantity of electricity. The battery used by the author is shown in Fig. 26. It has six revolving plates 26 inches in diameter, three long stationary plates, is easily changed, requires but little power to keep in motion, and will work in all kinds of weather. There is a form of static battery known as the Toepler, or Toepler-Holtz machine, and which has lately come into use, but the author has not had any experience with it.

Fig. 27 gives all of the electrodes needed in administering static electricity. The different methods of connecting up the Holtz machine, and the various forms of administration are fully given in Chapter V.

CHAPTER II

ELECTRO-PHYSIOLOGY.

The Human Body as a Conductor—Density.—By the term
density is understood the relative proportion of the strength
of current to the transverse section of the conductor through
which it passes. For example, if a certain strength of current,
say twenty milliamperes, is passed through a conductor one
inch square, and the same twenty milliamperes passed through
a conductor two inches square, the density would be four
times as great in the former as in the latter, because the current would be diffused through four times as much space in

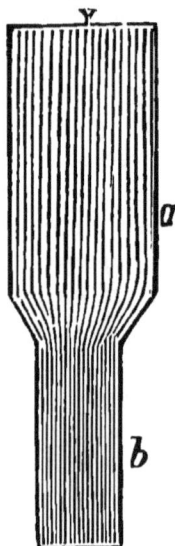

Fig. 28.

the latter case. Fig. 28, which is taken from Erb, serves
to illustrate this. One can easily see how much more
compact the threads representing the electric current are in
one part than they are in the other.

The integument, being a poor conductor, forms a large proportion of the resistance to the current in the medicinal use of
electricity. This resistance varies with circumstances. If the
skin is dry, its resistance is much greater than when well

moistened; therefore, when it is desirable to penetrate the skin so as to effect the underlying tissue, an electrode of some material that can be wet must be used. The conductivity of the skin will also be increased if this electrode is placed in position for a few minutes before the current is allowed to pass, so that the moisture will penetrate into the integument. I have seen a milliampere-meter needle pass from 50 to 120 without adding any more cells, in the treatment of fibroid tumors, when the current was turned on, the moment the abdominal electrode was placed in position. If this electrode is placed in position two or three minutes before the current

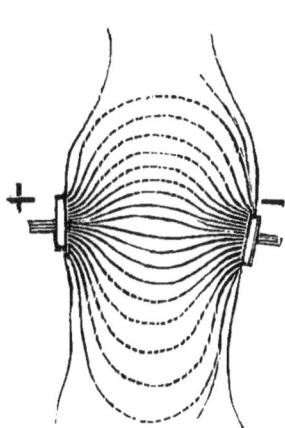 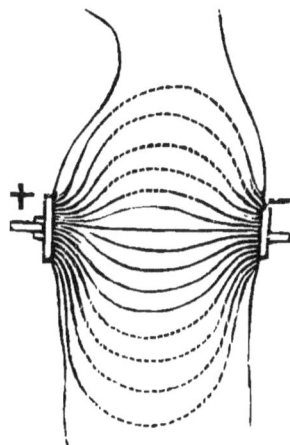

-Fig. 29.   Fig. 30.

is allowed to pass, so that the skin is well moistened, no such increase will occur. If the electrodes are wet in a solution of warm salt water the integument will absorb the moisture more readily and completely than when fresh water is used, and will, consequently, lessen the resistance.

The conductivity of the skin, like all other conductors, is greater the larger the surface acted upon; therefore, when electrodes of certain size and shape are not required, they should be as large as possible. On the other hand, if it is desired that the integument alone be acted upon, the greatest effect will be obtained by using small dry metal electrodes.

The interior of the body is a much better conductor than the integument. In order to understand thoroughly the application of electricity, we should have an idea of its distribution through the body. Let us imagine the electric current to be a number of parallel threads, the greater the strength of current the greater the number of threads, and the greater the number of threads in a given space the greater the density. If, now we place the positive electrode on the abdomen and the negative on the back, as in Fig. 29, we have a current passing from the positive to the negative. If Fig. 29 is studied, one will perceive that the greatest density of current is just beneath the electrodes. From this the current spreads in the interior of the body; the greatest density will be found in a straight line between the electrodes, because, in passing in a

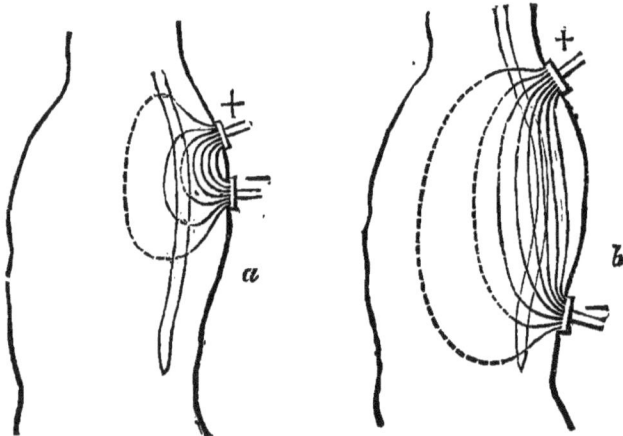

Fig. 31.   Fig. 32.

straight line, the current has a shorter distance to traverse, and, consequently, has less resistance; therefore, when it is our object to effect a lesion in the interior of the body, we should so place the electrodes that the diseased part will be in a direct line between them. If one electrode is larger than the other, as in Fig. 30, the density of the current will be seen to be much greater near the smaller; consequently when we use one as an active electrode and the other simply to connect the current on some indifferent part of the body, as in stimulating motor points, the inactive electrode should be as large as can conveniently be used. The above serves to illustrate the passage of a current when an electrode is placed on either side of the body; but it is important for us to understand the distribution of the current where both electrodes are placed on the same sides of the body, as in Fig. 31. It will here be seen that between the electrodes the greatest density of the current is just underneath the skin. If you were to compare Fig. 31 with Fig. 32 you will perceive in the latter that the electrodes are placed further apart, and, at the same time, the current penetrates the body to a much greater depth, and the greatest density of the current is some distance underneath the integument. Therefore, if we wish to penetrate deeply into the tissue with the electric current, such as sending a current lengthwise along the spinal cord, the electrodes should be placed some distance apart. The above has reference to the galvanic current. The faradic current does not penetrate so deeply. This explains the reason why a nerve deeply seated is more easily stimulated with the galvanic than the faradic current.

Electrotonus.—As the two poles of a galvanic battery have distinct actions which are sometimes just opposite to each other, it is essential that we study closely their individual actions.

The effect produced upon a nerve by the passage of a galvanic current through it is called electrotonus, and a nerve so affected is said to be in an electrotonic state. The positive pole is known as the anode, and that part of the nerve in the neighborhood of it and affected by it is said to be in an anelectrotonic state ; and, as the negative pole is known as the cathode, that part affected by it is in a catelectrotonic state. There is a neutral point where catelectrotonus meets anelectrotonus, and where the irritability is not changed. If a current of medium strength is used, the neutral point is midway between the two electrodes ; if weak, the neutral point is near the anode ; but if strong, near the cathode.

One very important point to remember in this connection is that a nerve in an anelectrotonic state has its irritability decreased, and a nerve in a catelectrotonic state has its irritability increased. It is for this reason that we employ the anode to relieve spasms by applying it over the affected nerve, which has its irritability increased, or for the relief of neuralgia by applying it over the hyperæsthetic nerve. It is for the same reason that we use the cathode when we wish to increase the irritability, as in some forms of paralysis. If the current be suddenly broken, it will be found that anelectrotonus suddenly changes to a state of increased irritability, and that catelectrotonus changes to a state of decreased irritability ; therefore, in the therapeutic use of electricity, where it is desired to derive benefit from the continuous effects of anelectrotonus and catelectrotonus, the current should be gradually raised to its maximum, and as gradually decreased.

There has been much controversy among physiologists about the continuous effect of anelectrotonus and catelectrotonus; but, while I do ·not believe in a system of therapeutics based entirely on the electrotonic theory, I am satisfied that, in certain pathological conditions, when the current is gradually increased and gradually decreased without any interruption, they do continue. The length of time varies with the strength and duration of the current used, the individual, and the character of the disease for which it is given ; but, as I have intimated, there is danger of carrying the polar action of the current too far.

I do not wish to be understood to say that the anode is always to be used to relieve pain, for the cathode will often be more efficacious. It has long been known by practical electrotherapeutists that the cathode is better to relieve muscular rheumatism than the anode. Still, I have often heard physicians recommend the use of the anode, and have also read the same in text-books, on the theory that the anode was to relieve pain ; but, as we at present understand, the cause of muscular rheumatism is due to a disproportion of the work done by it and the amount of nutrition it receives (that is, the nutrition it receives is too small for the amount of work required of it), we can easily understand why the cathode, with its increase of irritability and its greater power to increase the nutrition, is more prompt to cure the disease.

On the other hand, Dr. Mundé reported a case a few years ago of sciatica, due to pelvic inflammation and effusion, in which he had administered morphia for some time, and, becoming alarmed with the amount used, he finally concluded to try galvanism. He placed the cathode, armed with a ball electrode, on the sciatic nerve through the vagina, and the anode on the lower part of the thigh. Very much to his surprise, the

pains were immediately aggravated. The next day he called
Dr. Seguin in consultation. He advised placing the anode in
the vagina and the cathode on the thigh, and this was followed
by immediate relief, which continued for some time ; and five
or six treatments completely cured the sciatica. In this case,
had Dr. Mundé thought of the increased irritability caused by
the cathode, he certainly would not have placed it in so close
a proximity to the sore, inflamed, hyperæsthetic pelvis ; and
had he thought of the sedative effect of the anode, he would
not have hesitated which pole to have used in the vagina.

Motor Nerves.—A continuous flow of a galvanic current
through a motor nerve, unless it be very strong, does not produce contractions of the muscles which that nerve supplies.
If an electrode be placed over a motor nerve, and the current
gradually turned on, one cell at a time, no contractions will
follow; but if a current of sufficient strength be suddenly
closed over a motor nerve, contraction of the muscles supplied
by it will be produced. No matter how long the circuit remains closed, the muscles will immediately relax and continue
so as long as the uninterrupted current passes ; but under certain circumstances, when the current is broken, the same muscles will again contract. It was proven by Du Bois-Reymond
that the absolute amount of density of the current passing
through a motor nerve does not stimulate it, but that it
is stimulated by the change in the density from one moment
to another. It has also been shown that the more suddenly
the change of density is effected the stronger is the contraction. It is for this reason that the faradic current produces
such vigorous contractions of muscles when applied to the
motor nerves, for, as has been shown in electro-physics, the
current from a faradic battery is composed of several currents
of very short duration, the current being opened and closed
at every swing of the vibrator, and consequently the change
in the density is very sudden.

If a faradic battery is armed with a rheotome that vibrates
slowly, contraction will occur at every opening of the circuit,
for the opening of a faradic apparatus acts like the closing of
a galvanic current. When the rheotome is made to vibrate
rapidly, as with most of our faradic batteries, the changes of

density follow each other with such rapidity that the muscles do not have time to relax before another change of density causes them to contract again, and so on. It is for this reason that a muscle remains contracted under the stimulus of a faradic current all the time the current is passing, thus differing from the galvanic. If a galvanic current is rapidly interrupted with an automatic interruptor, it produces continuous contraction, the same as the faradic current.

With a galvanic battery the contraction will depend first upon the strength of the current used, for the change of density will be in exact proportion to the strength of the current, and second, the method used in interrupting the circuit. The strongest contraction being produced by interrupting the circuit with a metallic interrupter, such as a pole-changer, or with an interrupter in the handle of the electrode, because the interruptions will then be more sudden. The contractions will be much weaker when the electrode is taken off of the body and applied again, even when it is done rapidly, and if performed slowly no contractions are produced.

It is not necessary to entirely interrupt the current to produce muscular contraction. Any sudden increase in the strength of the current, provided the increase be sufficient in a unit of time, will produce muscular contraction. As we have seen, the galvanic current produces muscular contraction by its change in density, but these contractions differ materially with circumstances. A set of laws have been developed, by experimenting on frogs and other animals with the nerve laid bare; but, as they can not be verified in the living man, it is more important for us to understand the action of the motor nerves when stimulated with a galvanic current in a healthy individual, as we can draw many practical lessons from such knowledge.

Motor Points.—We should first understand what is meant by motor points. They are points on the surface of the body which correspond to that portion of the motor nerve which can be most easily stimulated by an electric current. They are the points where the nerves come nearest to the surface of the body. The motor point of a muscle is where the motor nerve enters the muscle. No physician can successfully and skillfully make an electrical diagnosis or treat many of the diseases of the nervous system without a thorough knowledge of the motor points, and this can only be acquired by carefully studying the charts and obtaining a practical idea by experimenting upon himself or some other individual. If we place one electrode, which should be small so as to increase the density, over the motor-point of a certain nerve, and the other, with a large electrode, on some indifferent part of the body, we will find that the contraction varies with the pole used and also with the opening and closing of the circuit. It will be found that when the cathode is placed over the motor point and the current is closed (designated as ca. cl.) the contraction will be much stronger than at any other closing or opening. Next to the cathodal closure contraction, ca. cl. c., is the anodal closure contraction (designated as an. cl. c.); that is, with the anode over the motor point and the circuit closed. Next to the anodal closure contraction, an. cl. c., comes the anodal opening contraction (designated an. o. c.); that is, with the anode over the motor point and the circuit broken. It is exceedingly difficult in health to produce a cathodal opening contraction, ca. o. c. This can be made more intelligible by experimenting on the peroneial nerve. You will find with a certain number of cells, say twelve, contraction will occur at ca. cl. but no contraction occurs at an. cl. or an. o. If you increase the number of cells to fifteen, you will find that the contracton is at ca. cl. will be stronger and you will also get a contraction at an. cl. but none at an. o. Again, increase the current to eighteen cells and you get a very strong contraction at ca. cl., which may be slightly tonic in character ; you will get a stronger contraction than the previous at an. cl., and also a contraction at an. o.

The following are the number of milliamperes, measured with a dead-beat milliampere-meter, it required to produce contraction of muscles by stimulating the healthy peroneial nerve covered with considerable adipose tissue in a man 26 years of age.

To produce slight contractions at

Ca. cl.	it	required	4	milliamperes.
An. cl.	"	"	10	" "
An. o.	"	"	24	" "

To produce a very strong contraction, which was tonic in character at

Ca. cl.	it required	22	milliamperes.	
An. cl.	" "	42	"	"
An. o.	" "	50	"	"

It is of the utmost importance that we fix these general proportions carefully in our minds, for in certain pathological conditions this normal formula is changed, and, consequently, becomes a great aid in making a diagnosis and prognosis. It also teaches us that, in cases of paralysis due to spinal disease, we will get the greatest contractions with ca. cl. ; that is, by placing the cathode over the motor points and causing the circuit to be closed preferably with a metallic interrupter in the handle of the electrode. There is another reason why we should use the cathode. It produces no pain on opening, while the anode produces nearly as much shock on opening as it does on closing.

We do not wish to convey the idea that it is impossible to produce continuous contraction with a constant flow of the galvanic current, but in order to produce such contractions more powerful currents are required than are ever used in this branch of electro-therapeutics.

Voluntary Muscles.—The voluntary muscles react to an electric current when applied to the muscle in the same manner as when it is applied to the motor nerve which enervates it. This is true so far as the reaction to the change of density is concerned, but in other respects they differ. When the current is applied to a muscle direct, that muscle alone contracts; but when the current is applied to a motor nerve, all of the muscles which it supplies contracts. Contractions are also less marked when the current is applied directly to the muscle, the best contractions being obtained when the electrodes are placed at either end of the muscle, or when one of the electrodes is applied over the point where the motor nerve enters it, which is the motor point of the muscle. Some muscles have more than one motor nerve entering them, consequently they have more than one motor point, and, in such cases, it is impossible to cause contractions of the whole muscle at one time by applying the current to one of them.

Muscles contract more easily when they are relaxed. This is important to know in the treatment of paralysis. A muscle also contracts much more readily when the patient wills it to contract. It has long been known as a therapeutic fact that when voluntary power is lost, improvement can be obtained by the patient trying to will the muscle to contract; and, when this is employed in conjunction with electricity, contractions will be stronger and improvement better than when the patient's will power remains indifferent to the treatment.

Involuntary Muscles.—Involuntary muscles react to the stimulus of an electric current in a different manner from the voluntary. Instead of contracting suddenly as soon as the current is closed, and relaxing again when a galvanic current is used, they contract slowly and continue contracting as long as the current is closed, and even after it is broken, with both the faradic and galvanic current, but the contractions are stronger when the galvanic currents is rapidly interrupted. The contractions of an involuntary muscle is made up of several contractions of the individual fibers of which it is composed. It has been proven that, when the current is applied directly to the viscera, the involuntary fibers of all of them can be made to contract. When it is applied on a living man we can produce the normal peristaltic action of all of the intestines. It is this power which gives electricity such a high standing as a curative agent in constipation. If a faradic current be so applied that a pole is at either end of the uterus, as with a double uterine electrode, contractions will take place and continue for some time after the application is discontinued. It is by this means that we are able to cause a large subinvoluted uterus to contract down to its normal size.

Sensory Nerves.—The sensory nerves are stimulated by a continuous flow of an electric current, either galvanic or faradic. With the galvanic current a sensation is felt with a weaker current at ca. cl., an. cl., and at an. o., than it is with a continuous flow. In this respect the sensory nerves follow precisely the same course as do the motor. When a weak galvanic current is applied to the skin the cathode is first felt and produces the stronger sensation all through the seance. At first a pricking sensation is felt at both poles; but, on increasing the strength of the current, this passes into a burning sensation, which is in proportion to the density of the current. Therefore, the larger the electrode the less the density and less the pain. The burning sensation is undoubtedly due, to a great extent, to the irritative effect of the chemicals which are set free at the surface of the electrodes. It will be remembered, in speaking of the electrotonic effect of the galvanic current, that the anode decreases the irritability. This is true in regard to its effect on the sensory nerves of the integument. After a strong current is allowed to pass for some time the burning will lessen at the anode, which is due to a slight anæsthesia of the integument underneath it. It is for this reason that a much stronger current can be borne in electrolysis when the anode is connected with the cutaneous electrode. The faradic battery produces a sticking sensation with every vibration of the rheotome. If the rheotome is made to vibrate rapidly, this sticking sensation becomes continuous, and, with the use of dry metallic electrodes with a strong current, becomes of an intensely burning character.

If the trunk of a sensory nerve be stimulated with an electric current, pain will be felt over the distribution of that nerve more marked with the faradic than with the galvanic current. This will be demonstrated by using an insulated electrode in the uterus. When the electrode is carried up into the cavity in the right fundus of the uterus it will stimulate the sciatic nerve, and pain will be felt all down the right leg and foot. If the electrode is carried to the left side, pain will be felt down the left leg and foot.

Nerves of Special Sense—Optic Nerve and Retina.—It has been demonstrated that this nerve reacts to the galvanic opening and closing by fixed laws. These laws show that ca. cl. and an. o. produce the same reaction, and that of ca. o. and an. cl. is the same. These reactions are always the same in the same person, but differ with different individuals, and, as I have proven in my experiments at the New York Ophthalmic Hospital,* in pathological conditions which, so far as the ophthalmoscope could demonstrate, were identical had altogether different reactions, it follows that no practical lesson can at present

---

* *North American Journal of Homœopathy,* December, 1886.

be drawn from studying these reactions either for diagnostic or therapeutical purposes. I, therefore, dismiss the subject.

Auditory Nerve.—What has been said regarding the optic nerve and retina is also practically true of the auditory nerve.

Gustatory Nerve.—If a mild galvanic current be applied to the tongue, a distinct metallic taste is noticed. If one electrode is placed on either side of the cheek, a distinct taste which differs materially is noticed on each side. The anode, which is strongest, produces a metallic alkaline, while the cathode a salty, biting taste. In applying electricity to the head, particularly if the anode is used on the head, a distinct metallic alkaline taste is observed. This is used as a guide, by some as an indicator, for the strength of current to be used, and will be referred to again in the department of "Special Therapeutics."

Olfactory Nerve.—This nerve is very difficult to stimulate with the galvanic current. If an electrode is applied directly to the schneiderian membrane an odor like phosphorus will be perceived.

The Brain.—While electrical experiments on the brain of animals, when that organ is exposed, have been of value to the physiologist, it is evident that they can not be of any use to the electro-therapeutist, as it is impossible to localize the current in the brain through the cranium.

The physiological action of the electric current on the brain of man is not well known. It is evident that the galvanic current penetrates into the substance of the brain, but it is not certain that the faradic current does. Contrary to what usually occurs, an. cl. produces greater stimulation of the cerebral cortex than ca. cl. If an electrode is placed on either temple and a strong current passed, the person experimented on will totter toward the side of the anode, but, on breaking the circuit, he will totter to a less extent toward the side of the cathode. The subjective sensation of tottering is much greater than what actually occurs. During the passage of the current there is an apparent movement of surrounding objects before the eyes, which seem to pass in an opposite direction to the tottering—that is, from the side of the anode to that of the cathode. If a very strong current is passed, oscillated movements of the eyeballs occur toward the side of the cathode. Galvanic vertigo is produced with much more readiness the greater the angle of the current is to the sagittal suture; but it is not impossible to produce vertigo when a current is applied longitudinally through the head, or with one electrode on the head and the other on the back or abdomen. In the latter case, the vertigo is much greater if the anode is the electrode applied to the head.

Spinal Cord.—There is still less known of the electrophysiology of the spinal cord of man than there is of the brain. It has, however, been shown that the galvanic current is capable of penetrating into the substance of the spinal cord; for, if a large, well-moistened electrode (attached to the cathode) be placed over the upper lumbar vertebræ and the current closed, ca. cl. contractions will take place in all the muscles supplied by the sciatic nerve. It is also evident that the spinal cord is affected from the brilliant therapeutic results which are obtained when the electric current is applied in pathological conditions; but those are chiefly due to the nutritive and electrolytic effect of the current, and will be referred to in the chapter on that subject

Electrical experiments on the sympathetic and pneumogastric nerves have hitherto failed to yield any information that can be used as a guide in practice.

Abdominal Viscera.—The effect produced on the abdominal viscera is due, first, to the action of the current on involuntary muscular fibers; second, to the nutritive effects of the current, and, third, to the electrolytic action. The first of these is most marked with the faradic current. With the second, it is difficult to say which current is most efficacious; while, with the third, the galvanic is, of course, the only one that has any electrolytic action. The effect of an electric current on involuntary muscles has been given. It has been shown how the normal peristaltic action of the stomach and intestines can be excited, and that the uterus can be made to contract upon itself. It has also been stated, on good authority, that by applying a strong faradic current over the abdomen, the gall-bladder can be made to contract, and that an enlarged spleen can be reduced to its normal size. The nutritive and electrolytic effects will be given in the following chapter.

CHAPTER III.

Electrolytic Action of the Galvanic Current.—The electrolytic action of the galvanic current may be divided into catalysis and electrolysis. This division is more practical than scientific. By the term catalysis we understand that process of disintegration and absorption which takes place when electrodes are placed on the surface of the body. By the term electrolysis we mean the phenomena which are produced when one or both of the electrodes are metallic and are introduced into the substance electrolyzed. We have here the same disintegration and absorbing process as in catalysis; but, added to it, we also have the local effect around the metal electrode. Catalysis is the interpolar effect produced by the passage of a galvanic current. Electrolysis has both the interpolar and polar effects of the current.

Catalysis.—When the electrodes are placed one on either side of a limb or the body and a galvanic current is passed from one to the other, we find that the liquid parts of the tissues are decomposed and broken up, the several atoms of which they are composed being rearranged to form new molecules. This is called the chemical action of the current, and can be illustrated by simply passing a current through water, which will be decomposed into hydrogen and oxygen. If a salt is added to the water, it will be decomposed ; the hydrogen and the alkalies will go to the negative pole, while the oxygen and acids will go to the positive pole. The same phenomena occur when the current is passed through a portion of the body : the fluids are decomposed, and the hydrogen and alkalies go to the negative and the acids and oxygen to the positive electrode.

Electrical Osmosis.—As has been stated, the electric current not only decomposes liquids and salts of the human body, but the products of decomposition actually go to the electrodes. This transportation of the different constituents of the body from one part to another is called electrical osmosis.

Cataphoric Action.—It has also been demonstrated that, if the positive pole is saturated with any detectable substance and a current allowed to pass, the same substance is detectable at the negative pole, thus proving that the current which runs from the positive to the negative pole has carried it through the tissue with itself. This is called the cataphoric action of the galvanic current.

Absorption.—After the electric current has decomposed the tissue, absorption takes place. This occurs both during and after the treatment, and varies with the treatment and character of the structure acted upon ; it is most marked in large cystic tumors. The natural power of absorption of the human body is greatly increased by the electric current. The various phenomena which take place are known as the catalytic action of the galvanic current, and to illustrate it more fully I have given a quotation from Erb's " Hand-Book of Electro-Therapeutics," viz.: " which causes dilatation of the blood-vessels and lymphatics, causing more ready circulation of the blood and nutritive fluids, and increased absorption : increased power of imbibition of the tissues, increased osmotic processes, and thus increase of volume (especially in the muscles) ; changes in the disassimilation and nutrition of the nerves on account of their stimulation or sedation ; changes in the molecular arrangement of the tissues caused by electrolytic processes ; finally, the consequences of the mechanical transport of fluids from one pole to the other."

Electrolysis.—What has just been said in regard to the osmotic, cataphoric and absorbing effects of the galvanic current in catalysis is also true in electrolysis, but we have also the additional, local effects of the electrodes introduced into the substance electrolyzed. These differ with the pole used and with the parts of tissue acted upon ; and, in case the positive pole is used, with the material of which it is composed. If an ordinary steel needle is used on the positive pole of the battery, the acids and other chemicals which accumulate around that pole will attack the steel, and it will at once become oxidized. The result will be, if the current is of sufficient strength and duration, that the whole needle will become dissolved by the action of the chemicals upon it; but with weak currents, and of short duration, the needle will become oxidized and stick to the surrounding tissue so that, in attempting to remove it, parts of the tissue will be torn away.

Therefore, with one or two exceptions, which will be mentioned in the department of Special Therapeutics, if the positive pole is used in electrolysis, it should be made either of gold or platinum.

If we introduce both positive and negative electrodes into living tissue, we find that the chemicals heretofore mentioned collect around each, and that these set up destructive action; the action around the negative being much greater than the one around the positive pole; therefore, when we desire to destroy fibrous growths, we should use the negative pole.

If we introduce the two electrodes of a galvanic battery into blood, we find that a clot is formed around each, but they differ materially in their make-up. The clot around the positive is small, hard, and closely adherent; while that around the negative is much larger, loosely attached, and is very soft—being filled with hydrogen. These differences are of importance in the treatment of aneurisms and bloody tumors of all kinds, and will be mentioned in the treatment of that subject in Special Therapeutics.

Chemical Galvano-Caustic Action.—If we place the two metal electrodes on the surface of living tissue, we find that an eschar is produced around each, and that the nature of these eschars is very different. The positive pole produces a dry, white, contractile eschar, which is hardening and drying and is markedly hæmostatic in its effects. It is owing to this special action that we use the positive pole as the internal electrode in hæmorrhagic fibroids of the uterus, or in those forms of metritis and endometritis which are accompanied by hæmorrhage or profuse leucorrhœa. The negative pole produces a soft eschar which is non-contractile, leaving a superficial cicatrix, and it is not hæmostatic in its effects, but the eschar is much larger than the one around the positive. These eschars are produced by the chemicals set free by the electric current and are termed the Galvano-Chemical Caustic.

The Increase of Nutrition by Electricity.—It has long been known that electricity has the power to increase the nutrition of the human body. In order to fully understand the methods in which the various forms of electricity accomplish this, we should divide the current into weak and strong. *Weak* currents are when there is not sufficient strength used to cause muscular contraction, and *strong* currents when contractions are produced.

A weak galvanic current passed through the body or a part of it increases the nutrition by increasing the supply of the nutritive fluids. When a weak faradic current is passed through the body, it is supposed that the constant opening and closing of the circuit acts in a similar manner to gentle tapping of the surface, which has long been recognized as a potent factor in increasing nutrition.

When strong currents are used so as to cause muscular contraction with galvanic, faradic, or static electricity, all alike increase nutrition of those muscles the same as any other exercise.

CHAPTER IV.

ELECTRO-DIAGNOSIS.

In order to understand the practical application of electro-diagnosis, it is necessary to first understand the laws which govern electro-excitability of the motor nerves and muscles; therefore, those who have not studied the second chapter of this volume should do so before going further.

Electro-diagnosis is based upon the knowledge which is derived from the abnormal reaction of nerves and muscles. In this chapter what is said regarding electro-diagnosis will be confined strictly to the motor-nerves. These changes are of two kinds: First, quantitative, and second, qualitative. Quantitative changes are simply an increase or diminution of the electrical excitability of motor nerves or muscles. An increase of excitability is made known by the nerve or muscle reacting to a current of less strength than it will in the normal state, or when with the same strength of current the contractions are much stronger.

The irritability is diminished when a current of much greater strength is required to cause contraction than is required in health ; or when with the same strength of current the contractions are much weaker. In quantitative increase or decrease of excitability, both nerves and muscles react in the same manner as in health ; that is, the muscles contract with a quick jerk, as in health.

The same laws that govern the reaction in health apply here also, that is, ca. cl. produces contraction with the weaker current, and an. cl. and an. o. follow in the order in which they are given. Qualitative changes: First, in the manner of contraction. Instead of the short, jerky contraction, as in health, it is long and drawn out : this is called a "modal" change. Second, the normal law of reaction to the galvanic current is changed; instead of ca. cl. producing contraction with the weakest current, an. cl. may do so, or the two may produce equal contraction with the same strength of current; or it may be that ca. cl. still leads, but the difference between them is less; instead of it being six milliamperes it may only be two or three milliamperes.

In some cases an. o. may produce contraction before an. cl. This is known as a "serial" change.

Reaction of Degeneration (designated R. D.).—As the reaction of the nerves and muscles in R. D. differ, it is necessary to consider them separately.

Changes in the Reaction of Nerves.—The change in the reaction of nerves is a quantitative diminution to both galvanic and faradic currents alike. This decrease is either rapid or slow, according to the acute or chronic character of the disease. In an acute disease, such as infantile paralysis, or in a severe traumatic lesion of a peripheral nerve, there is a rapid decrease in the electrical excitability from the first, which may become completely lost in one or two weeks; while, with a chronic or sub-acute affection, it may be months or years before it becomes lost.

The period during which a nerve remains unexcitable varies with the nature and extent of the lesion which causes it. There are rarely any qualitative changes in the reaction of the nerves, but a modal change may be observed—that is, the contraction, instead of being quick and jerky, may be long and drawn out or feeble.

Changes in the Reaction of Muscles.—With the muscles we have a more characteristic phenomenon. When a faradic current is applied direct to a healthy muscle, that muscle is made to contract by the stimulation of the intra-muscular nerve fibers—the muscular fibers not being stimulated by currents of such short duration as the faradic current is composed. It is, therefore, evident that the faradic current follows exactly the same course when applied to a muscle as it does when applied to a nerve. It has been shown that as the nerve degenerates it gradually loses its excitability to both galvanic and faradic currents, and, as the intra-muscular nerve fibers degenerate along with the trunk of the nerve, the faradic current will lose its exciting power when applied to one the same as to the other; but, as a nerve begins to degenerate at the point of injury and go toward the periphery, it follows that the intra-muscular nerve fibers will be the last to degenerate, and, consequently, it loses faradic irritability somewhat later than does the nerve, the difference varying with the character of the disease—being slight in acute, but longer in chronic diseases. Muscular fibers are capable of being stimulated with the galvanic current, irrespective of the nerve which supplies it, and its changes of reaction in R. D. are highly characteristic. The changes are: First, quantitative. This differs in acute and chronic affections; with the former there is at first a slight diminution in the excitability, but this is soon changed to an increase of excitability, and may even become so marked that only two or three cells will be needed to cause contractions that require ten or twelve in health or if the disease is unilateral on the healthy side. This over-excitability may persist for several weeks, when it gradually returns to normal or below normal, and in incurable cases the excitability may be lost. With chronic affections there is a gradual diminution of excitability from the first.

Second, qualitative changes. These consist, first, of a modal change, and should be carefully looked for, as it is often the only indication in slight affections that R. D. is present. The serial changes in R. D. are the most characteristic of any.

There is a form of partial R. D. which is only manifested by the qualitative changes in the muscles, the nerves reacting normally to the galvanic and faradaic currents, and the muscles also retaining their faradic excitability. This occurs when the injury or disease is very slight. The accompanying cut, Fig. 33, taken from Erb, illustrates the R. D. The ordinate, O, indicates the starting point or commencement of the disease; the galvanic and faradic excitability of the nerve is indicated by the lower line. The galvanic and faradic muscular excitability are indicated by separate lines. The wave of the galvano-muscular line shows the periods of qualitative changes. At the top are given the various histological changes that the nerve undergoes during the stages of degeneration and regeneration. Fig. 33 represents a mild case, Fig. 34 a more severe one, and Fig. 35 a severe, incurable case.

The Occurrence of Changes in Reaction and Their Practical

Application. First, Simple Increase and Diminution of Electro-excitability Without the Peculiar Phenomena which Indicate R. D.—A simple increase of irritability is found in cerebral

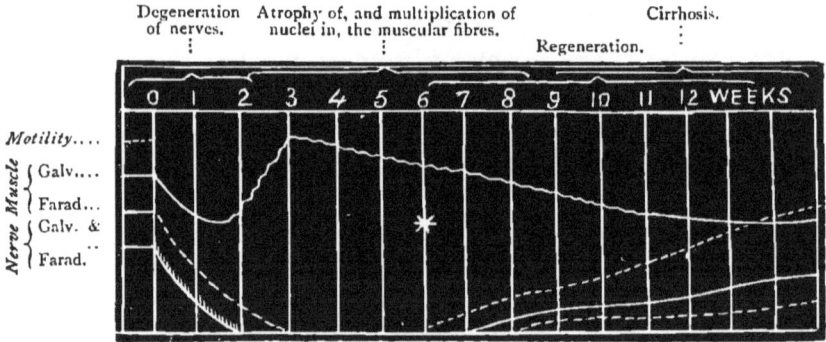

Degeneration   Atrophy of, and multiplication of   Cirrhosis.
of nerves.   nuclei in, the muscular fibres.
                                         Regeneration.

Motility....

Nerve Muscle { Galv.... / Farad... / Galv. & / Farad.

Fig. 33.—RECOVERY RAPID.

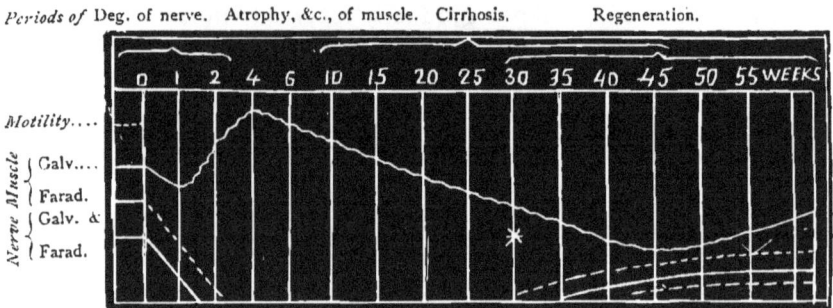

Periods of Deg. of nerve.   Atrophy, &c., of muscle.   Cirrhosis.   Regeneration.

Motility....

Nerve Muscle { Galv.... / Farad. / Galv. & / Farad.

Fig. 34.—RECOVERY SLOW.

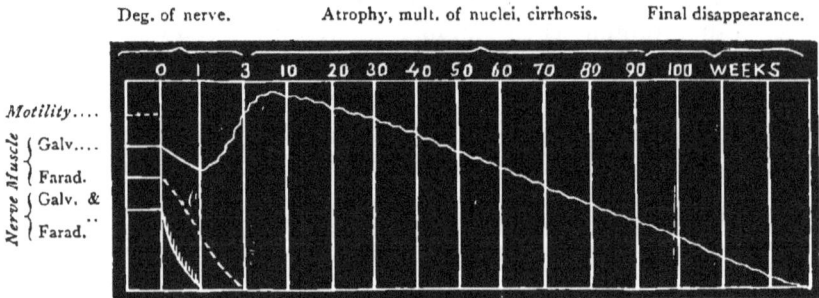

Deg. of nerve.   Atrophy, mult. of nuclei, cirrhosis.   Final disappearance.

Motility....

Nerve Muscle { Galv.... / Farad. / Galv. & / Farad.

Fig. 35.—NO RECOVERY.

paralysis of recent origin. It is also occasionally found in the first stage of locomotor ataxia some months before any other symptom appears to corroborate the diagnosis; but it is not of constant occurrence. The most striking example of increased irritability is found in certain forms of spasms, such as tetany.

Simple diminution of electrical excitability does not occur in cerebral paralysis except in those cases where the disease is of long standing and is associated with descending changes. It can, therefore, be considered as a fact that, if it exists to any degree, the disease is not purely cerebral. It also occurs in certain spinal affections, such as locomotor ataxia of long standing and diseases affecting the white matter of the cord. It is of very frequent occurrence in that class of diseases which come under the orthopædic surgeon's hands which are the result of spinal disease ; also, below the point of the lesion in transverse myelitis affecting the whole thickness of the cord, when due either to disease or traumatism, but the nerves and muscles under the direct influence of the diseased portion of the cord will show signs of R. D.

Muscles exhibit diminution of electro-excitability from dis-use, in chronic joint affections, also after a fracture or surgical operation which requires the limb to be kept in position for a long time. The practical importance of the occurrence of diminution, besides its contra-indicating purely cerebral disease, proves that some organic disease does exist, that it is not hysterical, and also that the patient is not shamming. The two latter bear the same relation to increased irritability.

Any of the above diseases may in some stages present a perfectly normal reaction. In this case we will have to com-pare the apparent disease or the disease feigned, and see if it is compatible with the reactions found. If it is hysterical, a good way to distinguish it from diseases that have normal reactions is, after causing several contractions of the muscles, remove the electrodes to some place where they do not in any way stimulate the part affected, and, after telling the patient the same contraction will be produced, suddenly close the circuit, and it will be found that very often the affected muscles will contract, which is proof that it is hysterical ; but its ab-sence does not prove that it is not hysterical.

Reaction of Degeneration.—Second, when R. D. is present, it indicates one of two things—that the motor nerve or the

gray matter of the anterior columns of the spine is the seat of
disease.   It further shows that there is more or less degenerative atrophy of one or both of them.   The most typical cases
of R. D., when the motor nerve is the seat of disease, is in
severe traumatic lesions causing section or rupture of it.   It is
well known in these cases that within from two to four days
the medullary sheath breaks up into granules, accompanied
with destruction of the axis cylinder; these afterward become
absorbed, leaving a homogeneous protoplasmic mass in the
sheath of Schwan.   After a shorter or longer time, according
to the nature and severity of the lesion, regeneration occurs.
If it has been slight, regeneration will be rapid and complete ;
if more severe, it will be slow and less complete, and if very
severe, no regeneration will take place.

Severe pressure, which causes a change in the nutrition of
the nerve, produces R. D. ; but mild pressure, not sufficient to
cause any change in the nutrition, may cause paralysis, but no
R. D.   Certain intersticial changes in the nerve will cause
R. D.   Facial paralysis is an example of the latter, when it is
of a rheumatic origin, which causes some parenchymatous
changes in the nerve, or when it is due to a blast of cold air
on the face, causing an effusion into the nerve sheath.   It
occurs in all spinal diseases which affect the gray anterior
column of the spine, whether acute, subacute, or chronic, such
as anterior poliomyelitis, chronic progressive bulbar paralysis
and amyotrophic lateral sclerosis.   It is present in spinal
hæmorrhage and in some forms of chronic myelitis, but only
when they affect the gray anterior columns.   The following
class of diseases may be excluded when R. D. is present:
First, diseases purely muscular, such as myositis or paralysis
from joint disease.   Second, those diseases which are distinctly cerebral ; and, third, disease of the white columns of
the cord alone.

It should be remembered that R. D. is sometimes present
when no paralysis exists, and that great atrophy of muscles
may exist without exhibition of R. D., and, also, that motion
is sometimes restored before electrical reactions return.   We
have seen that the presence of R. D. does not tell the particular disease from which the patient is suffering, but locates its

cause in the motor nerve or in the gray anterior columns. If the disease is due to the former, only those muscles supplied by it exhibit R. D.; but, if the latter are the seat of disease, we are very apt to have R. D. in muscles supplied by more than one nerve. This brings it down to a very few diseases, and these can be easily distinguished by other diagnostic points. After carefully studying the preceding pages, one can plainly see that valuable deductions can be drawn from abnormal reactions in regard to prognosis. It should be understood that in all grades of R. D., from a slight modal change to a complete loss of reaction, in the same disease and due to the same causes, the disease will be found to be of longer duration and in every respect more severe the later the stage of the disease in which it is found and the more complete is R. D.

Methods of Determining Qualitative and Quantitative Changes.—A well-practiced eye can determine very accurately any qualitative change, or it can be done by comparison with other muscles on the same individual, as the muscles in different parts of the body, so far as their qualitative reactions are concerned, are about the same. First examine a healthy muscle on some other part of the body and note its quick, jerky contractions. Then compare it with the diseased one, and, if there is a modal change, it will be found that the muscle rises to its maximum of contraction more slowly; it remains at the summit of contraction longer, and declines slower than does the healthy muscle. Again, cause the healthy muscle to contract at ca. cl., an. cl. and an. o., and carefully note the number of milliamperes it requires for each. Compare the same with the diseased muscle. If there is a serial change, it will be noticed that the difference in the number of milliamperes required to cause contraction is less according to the amount of R. D. that is present. It should be remembered that, when more than one motor nerve enters a muscle, and only one of the nerves is diseased, only that part of the muscle which is supplied by the diseased nerve will show changes in reaction. With the quantitative change one can never become so well practiced as to be able to determine it unless it is well marked, and it will not do to compare any two persons; therefore, we must take some more accurate method of comparison. I wish

here to point out a few of the fallacies which should be guard-
ed against in comparing different muscles. The inactive electrode should be placed in the median line, so that the resistance will be the same on both sides of the body; the active
electrode should be placed directly over the motor point, and
all other things which tend to make a difference in the resistance to the current should be carefully avoided. The circuit
should be made and broken with the same suddenness in each
case; the parts examined should be in an equal state of relaxa-
tion, and the patient's will power should not in any way assist
or retard the contraction. Over the sternum is the best place
to put the inactive electrode, as there it does not produce discomfort; is in the center of the body, and can be held by the
person examined. The active electrode should be well moistened every time it is changed, care being taken that the skin
is as well moistened in one place as in another, thus removing
all source of error. When the disease is unilateral, comparison
with the healthy side is easy, and very slight quantitative
changes can be ascertained; but, when the disease is bilateral,
we have to seek some other means of comparison. Erb first instituted comparison of different parts of the body. He found
that there was quite a uniformity between the reaction of the
facial, spinal accessory, ulnar and peroneal nerves in a
healthy individual, and that any great departure from this
normal relationship was the result of some pathological condition. Erb's method was to place a galvanometer in the circuit, and, with a certain number of cells, note the deflection
of the needle.

Since Erb's experiments, we have come in possession of the
milliampere-meter, which accurately measures the current. The
writer has instituted another method of making a comparison
of these different nerves. His method is, first observe all the
details just given, to avoid any source of error, although it is
not so essential as it is with Erb's method. A larger number
of cells is turned on than is needed to produce the required
contraction. A water rheostat, capable of very fine adjust-
ment, and a dead-beat milliampere-meter, with a registering
scale graded to one-fifth of a milliampere, is placed in the circuit. The current is then graduated by means of the water

rheostat until it is sufficient to produce slight contractions. The circuit is then closed for ten seconds, when the needle will have become quiet; the register is then taken. No matter if the needle should become quiet before the ten seconds expire, do not read the register before, as the longer the circuit is closed the more the current will measure. It is, therefore, necessary to have some stated time for the current to pass before it is closed and the register taken. With the instrument I use (Waite & Bartlett's dead-beat milliampere-meter) the needle becomes quiet in ten seconds under any circumstances ; therefore, I have chosen this time. A large, flat electrode, well moistened, is placed on the sternum, and, with the active electrode on the motor points of the spinal accessory, facial, ulnar and peroneal nerves, successively stimulate them. The following are the figures obtained by so testing two healthy men :

	First case.	Second case.
Right facial,	5 milliam.	2.8 milliam.
Left "	5 "	2.8 "
Right spinal acc.	3 "	.8 "
Left " "	2.6 "	.8 "
Right ulnar,	2.6 "	1.5 "
Left "	2.2 "	2 "
Right peroneal,	4 "	1.8 "
Left "	4 "	1.6 "

Out of ten healthy cases in which I applied the same test, those given above represent the extreme number of milliamperes required to cause muscular contraction, the first requiring the largest number and the second the smallest number of milliamperes. If the above figures are carefully studied, one will immediately recognize the relative proportion in the number of milliamperes required to cause contraction, and any great deviation from this relative proportion can be considered as an evidence of increased or diminished excitability. In making an electro-diagnostic test with the faradic current, I use the Du Bois-Reymond coil, with the scale that marks the distance ; the secondary coil is shoved over the primary, graduated in millimetres (see Chap. I.). In the following table, when it is marked four millimetres or five millimetres, it means that

the secondary coil is shoved on the primary that distance. In
these tests the secondary coil is always used.   As there is no
standard strength of faradic batteries, these figures are only
good for my instrument ; but they illustrate the relative proportion just the same, and, as with the test with the galvanic
battery, any great deviation from this relative proportion indicates a pathological condition.   The following figures were
obtained by testing two healthy men :

	First case.	Second case.
Right facial,	45 millimetres.	28 millimetres.
Left "	35 "	26 "
Right spinal acc.	20 "	13 "
Left " "	17 "	10 "
Right ulnar,	20 "	15 "
Left "	20 "	19 "
Right peroneal,	25 "	18 "
Left "	25 "	15 "

I do not give the above figures, obtained with either the galvanic or faradic currents, as a perfect standard for healthy reaction, for, with our present method of examination, it would
be impossible to obtain such a standard, and any slight deviation should not be looked upon as indicating disease, but only
when there is a marked change should they be considered abnormal.

CHAPTER V.

GENERAL THERAPEUTICS.

In the past electricity has been looked upon by some physicians as being a great medicinal agent, while others discarded it as useless ; but at present the medical profession as a whole believe more or less in its therapeutic value.

There are various reasons why electricity has not advanced much faster and held a higher position in therapeutics at an earlier day. One cause was the defective apparatus, another that the earlier experimenters thought one form of battery was as good as another, and would use the galvanic, faradic, or static according to their fancy, not realizing that each had its special sphere of action.

The failure on the part of physicians to obtain results which had been expected, from the reports of marvelous cures by both untrustworthy and incompetent observers, had no little share in prejudicing them against electricity. Being an agent of wonderful demonstration, electricity furnished such a good opportunity for charlatans that it has been one of their greatest fields of action, and, consequently, they have done much to throw it into ridicule in the minds of the people. Nearly all these obstacles have passed away. The time for quackery to produce any effect on the human mind is both limited and transient, and, while the apparatus is far from being perfect, it is about all that can be desired for medical use. The indications for the use of the different currents are very well established and fast developing. The subject is no longer confined to charlatans or an inferior class of physicians, but is being taken up and thoroughly investigated by original thinkers and scientific men, which is all any subject of merit needs to guarantee it a success. I do not wish to be understood to say that all former observers were unscientific, for some of them ranked among the highest authorities, but the majority did not.

Electricity is not a cure-all, but has its special sphere of action and indications the same as any other remedy, and the more closely these indications are studied and the treatment applied accordingly, the surer will be the success.

It has also had its successes and failures. Many physicians will try it once, and if it fails they discard it altogether for that disease, claiming that, if it failed in one, it would in all ; yet that same physician would try salicylic acid in rheumatism again and again, if it failed several times, which it would be sure to do if he gave it much of a trial. Electricity should not be discarded without a trial in at least a second case of locomotor ataxia, or any other disease, because it failed in the first, any more than any other remedy should under the same circumstances.

General Application.—In beginning with a case, you will often have to deal with the prejudice and nervous fear of the patient. It is your duty at all times to try and cause but little pain. This I know is sometimes impossible, but it should be a rule not to hurt or distress your patient in any way until they have become acquainted with the method and acquired confidence in you, when you will have no difficulty in getting them to bear the necessary burning or other disagreeable sensations. Never allow the patient to get in such a state of mind that they dread the application, for it will certainly hinder, if it does not totally destroy, your success. In order to receive the most benefit from the application of electricity, the patient should lie down for from thirty minutes to one hour after each treatment. It is particularly necessary that a patient should lie down after the general administration of electricity, or she will lose fully half of the benefits that would otherwise be attained.

The patient will probably feel a quiet, drowsy sensation, and will generally fall asleep, if she has a chance.

Labile and Stable* Applications.—When the electrodes are held steadily in one place the application is said to be stable ; when one or both is moved over the body the application is

---

* This in text-books is spelled *stabile*, which is the Latin, but as the word *labile* is English I have concluded to Anglicize *stable*, and spell it according to Webster.

said to be labile—this produces a vigorous irritation similar to the interruption of a current, but not so marked.

Motor Points.—These have been considered in the chapter on electro-physiology. I speak of them here to remind the reader of their importance and give the charts illustrating them.

General Consideration and Application of the Faradic Current.—It is a well-established fact that the secondary coil of a faradic battery differs materially in its effects according to the length and size of the wire of which it is constructed. When it is the desire to relieve pain, the secondary coil should be made of long, thin wire, not larger than 34 to 36; while, on the other hand, if we desire to cause muscular contractions, the wire should be short and thick, about 22 (see Chapter I.). This should be constantly borne in mind. If a coil composed of large and short wire were to be used in a painful cellulitis it would aggravate all the symptoms, while the finer coil would cause immediate relief. If you wish to produce muscular contractions of either involuntary or voluntary muscles, it can be done more effectually and with less pain with a coil composed of short, thick wire.

General Faradization.—By this method of administering electricity, we bring the whole of the body under its influence. The patient should have on a loose wrapper so that the physician can get at all points of the body. The feet are immersed in a tub of warm water ; or, if this is inconvenient, they can be placed on a foot electrode, but this is liable to cause disagreeable sensations. The negative pole of the battery should be connected with the foot-tub or plate. The operator begins by applying the positive pole to the head, forehead, neck, back, and so on all over the body. Those parts which are particularly sensitive, such as the forehead, points of the scapula, etc., will best be treated with the hand of the operator in place of the electrodes. All points that are sensitive should be studiously avoided, or treated with a current not strong enough to produce a disagreeable sensation. General faradization is a tonic in effect, and has a very wide range of usefulness, although I cannot but feel its importance has been over-estimated by its originators (Beard and Rockwell).

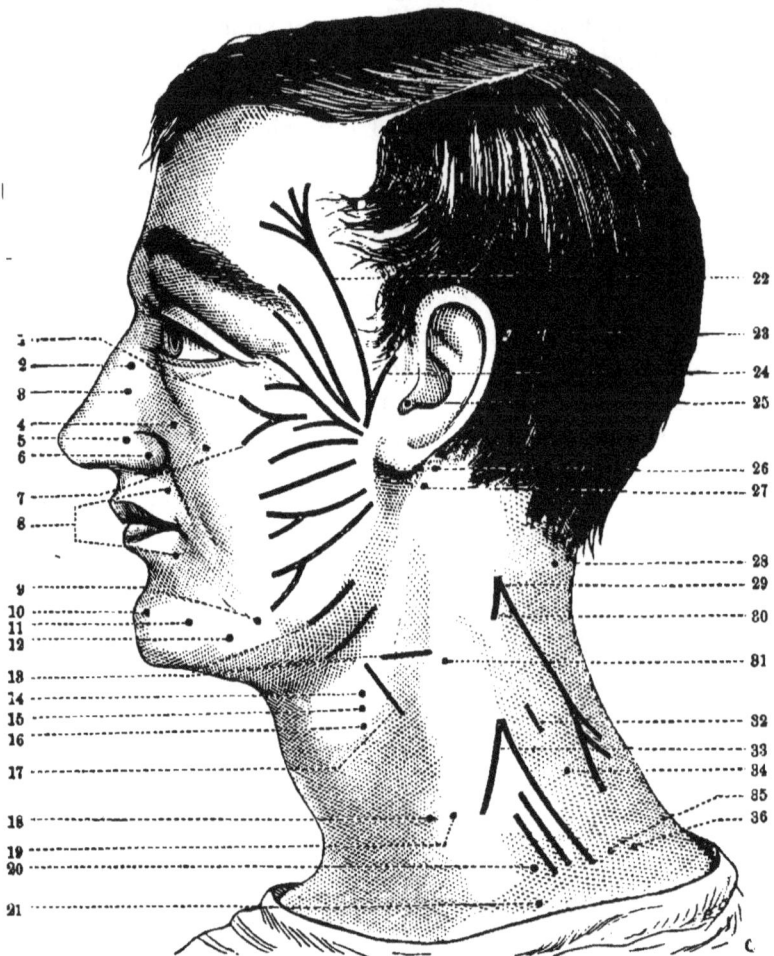

Fig. 36.—A Diagram of the Motor Points of the Face, showing the Position of the Electrodes during Electrization of Special Muscles and Nerves. The Anode is supposed to be Placed in the Mastoid Fossa, and the Cathode upon the Part indicated in the Diagram.

1, m. orbicularis palpebrarum; 2, m. pyramidalis nasi; 3, m. lev. lab. sup. et nasi; 4, m. lev. lab. sup. propr.; 5, 6, m. dilator naris; 7, m. zygomatic major; 8, m. orbicularis oris; 9, n. branch for levator menti; 10, m. levator menti; 11, m. quadratus menti; 12, m. triangularis menti; 13, nerves—subcutaneous, of neck; 14, m. sterno-hyoid; 15, m. omo-hyoid; 16, m. sterno-thyroid; 17, n. branch for platysma; 18, m. sterno-hyoid; 19, m. omo-hyoid; 20, 21, nerves to pectoral muscles; 22, m. occipito-frontalis (ant. belly); 23, m. occipito-frontalis (post. belly); 24, m. retrahens and attollens aurem; 25, nerve—facial; 26, m. stylo-hyoid; 27, m. digastric; 28, m. splenius capitis; 29, nerve—external branch of spinal accessory; 30, m. sterno-mastoid; 31, m. sterno-mastoid; 32, m. levator anguli scapulæ; 33, nerve—phrenic; 34, nerve—posterior thoracic; 35, m. serratus magnus; 36, nerves of the axillary space. In this text m. = muscle; n. = nerve.

Its indications will be given in the special diseases in which it is useful, in the department of Special Therapeutics.

Galvano-Faradization.—By this method of electrolization we utilize both galvanic and faradic currents at the same time. The positive pole of the galvanic battery should be attached by means of a wire to the negative pole of a secondary faradic coil, and the rheophores attached to the two remaining poles. The same number of cells should be employed as would be used if the faradic current was not in the circuit, the strength

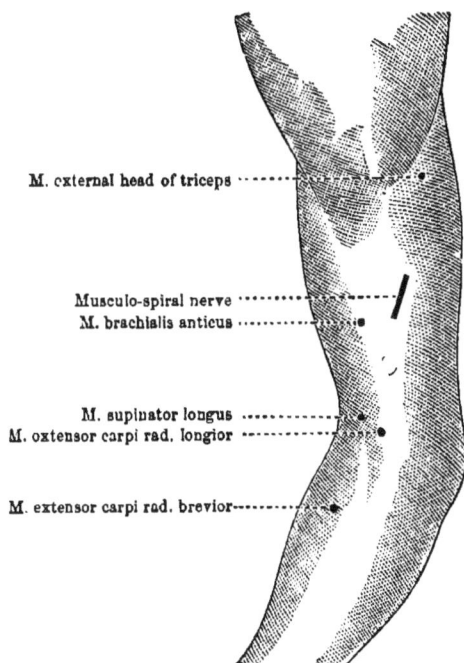

M. external head of triceps ········

Musculo-spiral nerve ·········
M. brachialis anticus ···········

M. supinator longus ···········
M. extensor carpi rad. longior ·········

M. extensor carpi rad. brevior········

Fig. 37.—THE MOTOR POINTS ON THE OUTER ASPECT OF THE ARM.

of the faradic current being governed by the amount of irritation desired.

Galvano-faradization has a strong irritating effect, and should be used in those cases where we wish to get the deep penetrating effect of the galvanic and the irritating effect of the faradic, such as stimulating the intestines in constipation. The electrode attached to the negative pole of the galvanic current has the greatest irritating effect.

Application of the Galvanic Current—Central Galvaniza-tion.—This is the method of administering the galvanic current so that all the central nervous system, the brain, sympathetic nerves and spinal cord are brought under its influence. The negative-electrode is a large flat sponge, and held firmly over the solar plexus. After the hair is wet, the positive pole is first pressed firmly on the head, then labile applications are

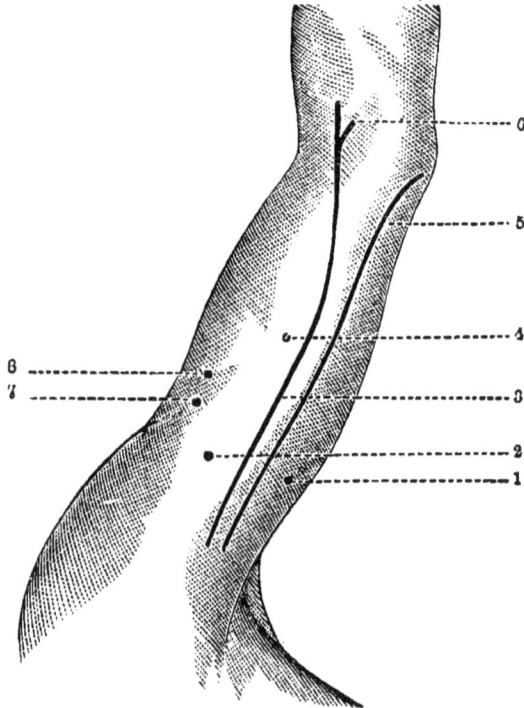

Fig. 38.—THE MOTOR POINTS ON THE INNER SIDE OF THE ARM.

1, m. internal head of triceps; 2, musculo-cutaneous nerve; 3, median nerve; 4, m. coraco-brachialis; 5, ulnar nerve; 6, branch of median nerve for pronator radii teres; 7, musculo-cutaneous nerve; 8, m. biceps flexor cubiti.

made from the top of the head to the occiput, from the mastoid process to the sternum and from the occiput down the whole length of the spinal column. The application should continue about five minutes, and should be given sufficiently strong to produce a metallic taste in the mouth when it is applied to the head and neck. Vertigo should be avoided, if possible. From eight to fifteen cells will be required.

General Electrolization.—This is the method of giving electricity so that all of the superficial and some of the deep

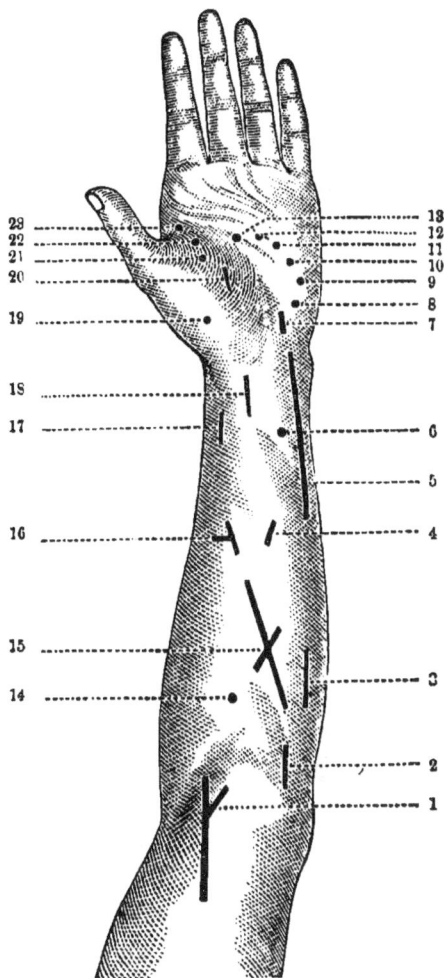

Fig. 39.—THE MOTOR POINTS ON THE FLEXOR (ANTERIOR) ASPECT OF THE FOREARM.

1, median nerve and branch to m. pronator radii teres; 2, m. palmaris longus; 3, m. flexor carpi ulnaris; 4, m. flexor sublim. digit.; 5, ulnar nerve; 6, m. flex. sublim. dig.; 7, volar branch of the ulnar nerve; 8, m. palmaris brevis; 9, m. abductor min. digit.; 10, m. flexor min. digit.; 11, m. opponens min. digit.; 12, 13, m. lumbricales; 14, m. flexor carpi radialis; 15, m. flexor profund. digitorum; 16, m. flexor sublim. digitorum; 17, m. flex. longus pollicis; 18, median nerve; 19, m. opponens pollicis; 20, m. abductor pollicis; 21, m. flexor brevis pollicis; 22, m. adductor pollicis; 23, m. first lumbricalis.

muscles of the body are made to contract, thus giving exercise to the whole body. Either current may be employed, but I prefer the galvanic, as it is less irritating, and consequently better borne. I seat the patient on a large electrode, which is so bent that it presses against the lower end of the spine; to this is attached the positive pole. The negative pole is

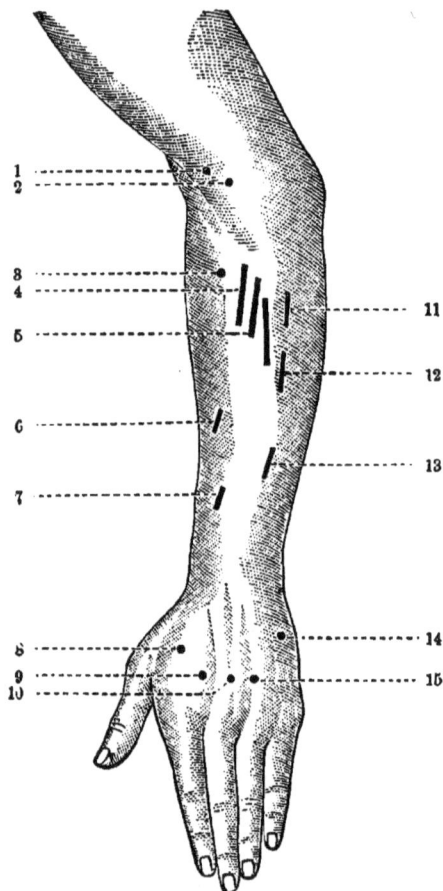

Fig. 40.—The Motor Points on the Extensor (Posterior) Aspect of the Forearm.

1, m. supinator longus; 2, m. extensor carpi rad. longior; 3, m. extensor carpi rad. brevior; 4, 5. m. extensor communis digitorum; 6, m. extensor ossis. met. pol.; 7, m. extensor primi. internod. pol.; 8, m. first dorsal interosseous; 9, m. second dorsal interosseous; 10, m. third dorsal interosseous; 11, m. extensor carpi ulnaris; 12, m. extensor min. digiti; 13, m. extensor secund. internod. pol.; 14, m. abduct. min. digiti; 15, m. fourth dorsal interosseous.

attached to an interrupting handle with a small-sized electrode, and this is used to go over the motor points of the body. From five to ten vigorous contractions should be produced

from each motor point. This method of administering electricity is a general nutritive tonic. It gives tone to the
patient and at the same time develops the muscles in a
remarkable manner. I have seen limbs increased half an
inch in circumference in a few weeks by such treatment,
with a corresponding increase in the strength, vigor, and en-
durance of the patient.

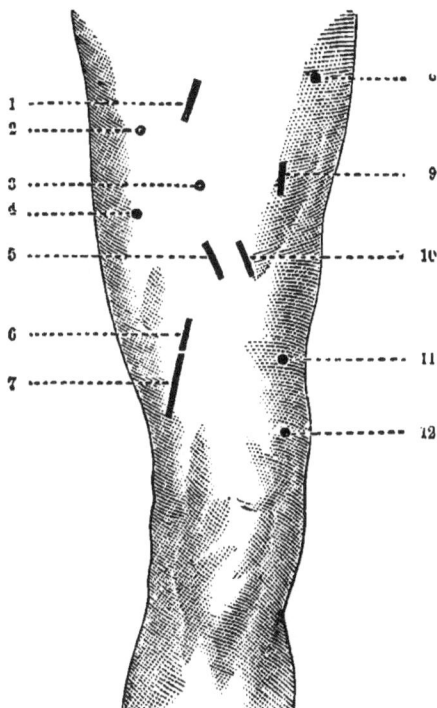

Fig. 41.—The Motor Points on the Anterior Aspect of the Thigh.

1, crural nerve; 2, obturator nerve; 3, sartorius muscle; 4, adductor longus muscle; 5, branch of the
anterior crural nerve for the quadriceps extensor muscle; 6, the quadriceps muscle; 7, branch
of anterior crural nerve to the vastus internus muscle; 8, tensor vaginæ femoris muscle (supplied by the superior gluteal nerve); 9, external cutaneous branch of anterior crural nerve;
10, rectus femoris muscle; 11, 12, vastus externus muscle.

Direction of Current.—Ascending and Descending.—When
the positive pole is placed on the spine or on a nerve near
where it is given off, and the negative on a peripheral portion,
—or when the positive pole is placed on the occiput, and the
negative on the lower part of the spine, the current is said to
be descending, for it runs from the positive down to the nega-

tive. It is also called direct, as the electric current runs in the same direction as the nerve current.

When the position of the electrodes are reversed, the current is said to be ascending or indirect.

Static Application.—In regard to the method of charging a static battery, I refer the reader to Chapter I.

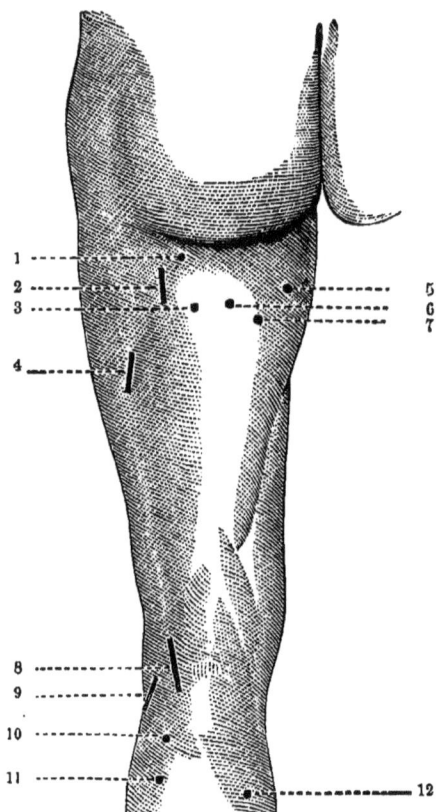

Fig. 42.—The Motor Points on the Posterior Aspect of the Thigh..

1, branch of the inferior gluteal nerve to the gluteus maximus muscle; 2, sciatic nerve; 3, long head of biceps muscle; 4, short head of biceps muscle; 5, adductor magnus muscle; 6, semitendinosus muscle; 7, semi-membranosus muscle; 8, tibial nerve; 9, peroneal nerve; 10, external head of gastrocnemius muscle; 11, soleus muscle; 12, internal head of gastrocnemius muscle.

The methods of administering static electricity are numerous and require careful consideration. The severity and length of the spark can be modified by the distance the connecting rods of the machine are drawn apart and the rapidity with which the plates are made to revolve. The greater the speed and the farther the connecting rods are drawn apart, the greater the strength and the length of the spark. The spark is also modified by the method of administering, by the size of

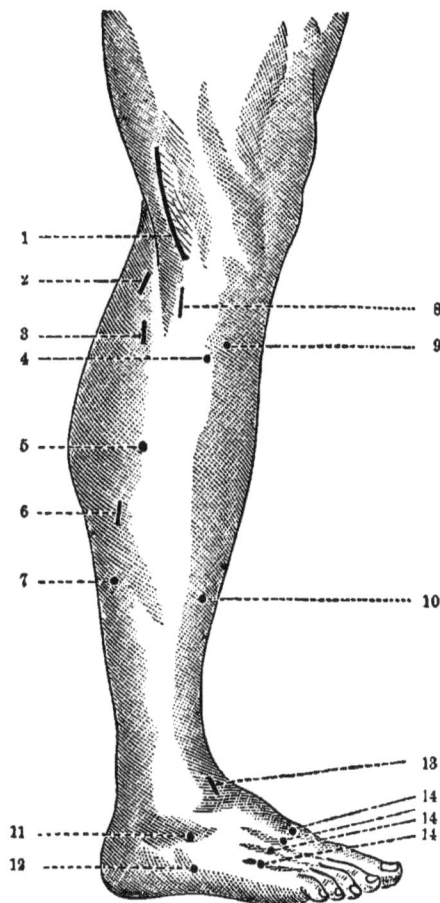

Fig. 43.—THE MOTOR POINTS ON THE OUTER ASPECT OF THE LEG.

1, peroneal nerve; 2, external head of gastrocnemius muscle; 3, soleus muscle; 4, extensor communis digitorum muscle; 5, peroneus brevis muscle; 6, soleus muscle; 7, flexor longus pollicis; 8, peroneus longus muscle; 9, tibialis anticus muscle; 10, extensor longus pollicis muscle; 11, extensor brevis digitorum muscle; 12, abductor minimi digiti muscle; 13, deep branch of the peroneal nerve to the extensor brevis digitorum muscle; 14, 14, 14, dorsal interossei muscles.

the ball electrode—the large brass electrode producing a much stronger spark than the small one—and the resistance in the circuit (the brass electrode being a much better conductor than the wooden one, yields a much stronger spark). The methods of administration will be given here, but the therapeutic indications will be found in the chapter on Special Therapeutics. The operator should be very careful not to unnecessarily shock the patient. A ring electrode should

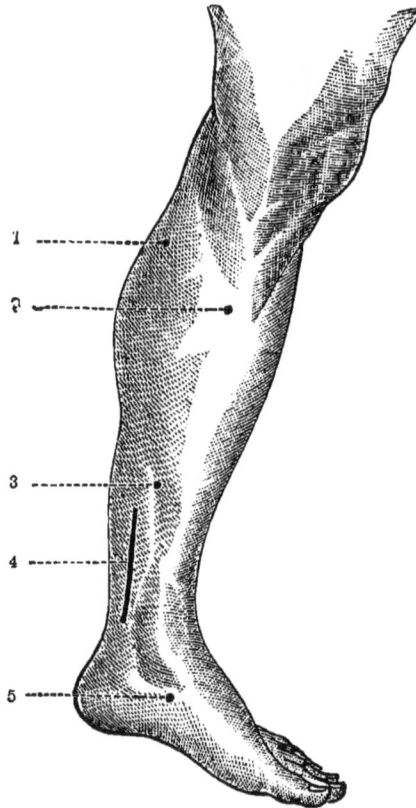

Fig. 44.—THE MOTOR POINTS ON THE INNER ASPECT OF THE LEG.

1, internal head of gastrocnemius muscle; 2, soleus muscle; 3, flexor communis digitorum muscle; 4, posterior tibial nerve; 5, abductor pollicis muscle.

always be used to keep the chain, which causes such shocks, from touching the patient.

The methods of administration are: Static Insulation. Static Breeze. Indirect Spark. Static Electro-massage. Direct Spark. Static Induction Current.

Static Insulation.—This is accomplished by placing the patient on an insulated platform, as in Fig. 45. The stool is connected with one of the poles of the battery, the connecting bars are drawn wide apart, and the opposite pole is connected with the ground, so as to draw off the electricity collecting on it.

Static insulation is a very pleasant way of taking static electricity. The patient is charged with electricity, and it passes off so gradually that it is not noticeable. Care should be taken not to allow any object to come near the patient, or a spark will pass and cause a shock. The hair is deflected and becomes erect, unless oiled. A slight tendency to perspira-

Fig. 45.

tion is produced, and the patient feels a quiet, soothing sensation.

Indirect Spark.—Fig. 46. The patient is placed on an insulated platform which is connected with one pole of the battery, while the other pole is grounded the same as for static insulation. Another electrode, connected with the ground, or preferably a gas pipe, armed with the large brass ball electrode, is placed a certain distance from the patient (from two to four inches) and a spark passes between them. When a static battery is charged, set in motion, and the connecting rods drawn

a short distance apart, a spark is seen to go from one to the
other. This is simply an equalizing of the potentials, one

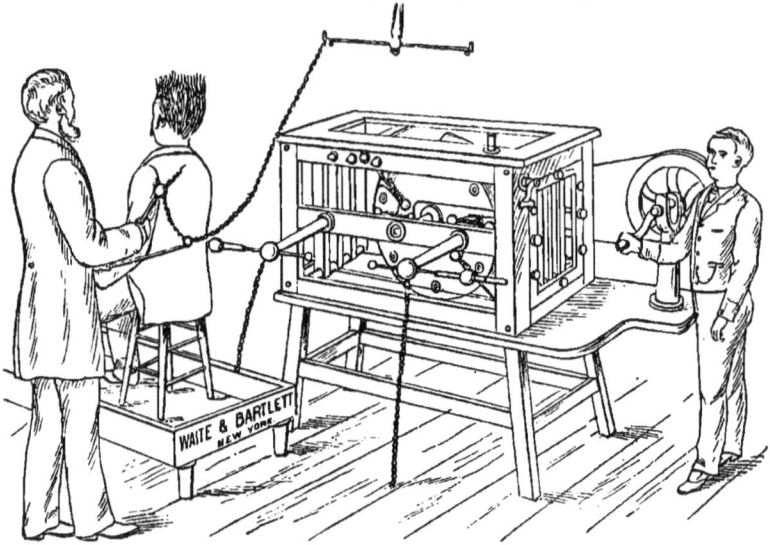

Fig. 46.

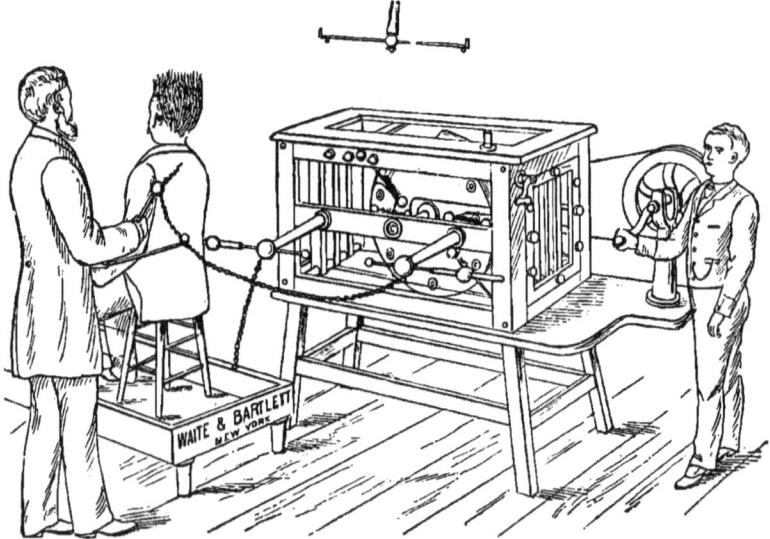

Fig. 47.

going from the higher to the lower, the other from the lower to
the higher, until they meet.

The same phenomena occur when a patient is on an insulated stool and a spark drawn from her—the current equalizes its potentials through the body. With the indirect applica-

Fig. 48.

Fig. 49.

tion the earth is also intervened, and the electricity has to take an indirect course through it.

Direct Spark.—Fig. 47. The direct spark is the same as the indirect, with the exception that the earth is not intervened between. The patient is placed on an insulated platform the same as before, but the opposite pole is connected direct with

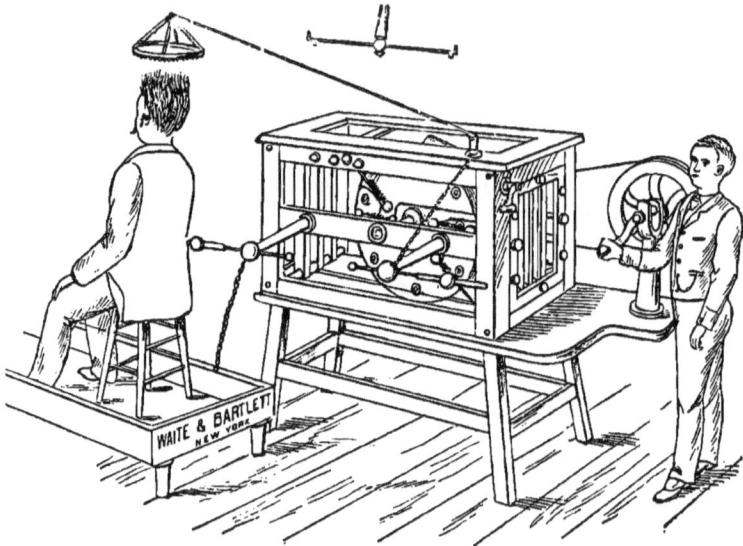

Fig. 50.

Fig. 51.

the ball electrode. This is a much stronger application than the indirect spark, and produces greater effects.

Both the direct and indirect spark produce muscular contraction, and both produce irritation of the skin, leaving little round raised spots known as " *wheals*," and, if the treatment has been severe, burning, which continues for some time after. If the treatment is on the face or any exposed surface, the part should be rubbed gently with a soft handkerchief.

Static Breeze.—This method of administering static electricity differs from those just described, in that instead of the electrode being a round ball it is flat, with many projecting points which divide the electric discharge into a fine spray. The electric breeze may be indirect, as in Fig. 48, or direct, as in Fig. 49. The electric head bath, as represented in Fig. 50, is simply the static breeze applied to the head.

Fig. 52.

With all the above varieties of administering static electricity, Leyden jars may be used or not. If it is desired to get a heavy shock, the jars should be used ; the size of the jars to be governed by the severity of the shock desired. The spark is much weaker when the jars are not used.

Static Electro-massage.—This form of administering static electricity does not require the use of the insulated platform. The patient is seated on a chair, with the feet on an ordinary foot-plate which is connected with one pole of the battery, as in Fig. 51. The roller electrode is attached to the other pole. The connecting rods at the top are now placed in contact with each other, and the machine set in motion. The roller is applied to the part that is to be treated, and the connecting rods very gradually drawn apart, when a pricking sensation will be felt underneath the roller. The current is stronger the further the connecting rods are drawn apart. Care should be taken not to draw them suddenly apart, or you will shock the patient. The best way is to separate them with a kind of twisting mo. tion. This is generally given without the Leyden jars. If it is desired to give a particularly strong treatment, the jars can be used, but the connecting rod which connects their external coverings must be removed.

All of the preceding methods of administering static electricity do not require that the clothing be removed, unless it be wet, as the electro-motive force is sufficient to carry the electricity through it.

Static Induction.—This is practically about the same as a treatment with an ordinary faradic battery ; but it possesses no advantages over that, and is certainly more troublesome to give. The part to be treated is to be laid bare the same as for treatment with dynamic currents. A pair of Leyden jars are placed on the machine as in Fig. 52. The rod connecting their outer covering is removed, and two rheophores are attached to the hooks which serve to hold them. The upper connectors are first approximated and then drawn apart the same as in static electro-massage. The strength of current is dependent upon the distance the connectors are withdrawn and the size of the Leyden jars used.

CHAPTER VI.

Diseases of the Brain.—In treating diseases of the brain, we may have three objects in view. First, to get the catalytic and nutritive effects of the current by direct electrolization of the head ; second, to effect the nutrition of the brain through reflex action by electrolizing the sympathetic and cutaneous nerves ; and third, to stimulate the nutrition of the muscles or other tissues which have been affected by the disease by labile or interrupted applications directly to those parts.

In giving electricity to the head we should be careful never to allow the current to be interrupted, but always stable or labile. While slight vertigo is not to be feared, we should be careful not to produce too great a sensation of vertigo, for it is liable to leave a sickening sensation which may continue for a long time after the treatment. For this reason you will have to give much weaker currents transversely through the head than longitudinally, as the greater the angle of the current, or point of application, to the sagittal suture, the more readily will vertigo be produced.

Cerebral Anæmia.—Experience has, I believe, proved that, when the positive electrode is placed over the occiput and the negative on the forehead electricity causes an increase of blood supply to the head. This, then, should be one of the principal methods of treating anæmia.

A current that is barely perceptible should at first be used, and then gradually increased with each treatment until quite strong currents are given. Each application should occupy from three to four minutes.

Central galvanization is often beneficial. The most effective method is to combine the two. Give the former for two minutes, and then follow with central galvanization. The current may also be given through the head transversely and obliquely.

A faradic current passed longitudinally through the head will often be beneficial.

Cerebral Hyperæmia.—Here we wish to get just the opposite effect that we did in anæmia, consequently we reverse the positions of the electrodes; that is, place the negative at the back of the neck and the positive on the forehead and gradually increase the current, as with cerebral anæmia. As an adjunct to this form of treatment, cutaneous faradization of the limbs and trunk, which causes hyperæmia of the integument, relieves the congestion of the internal organs and brain.

Cerebral Hæmorrhage.—In treating cerebral apoplexy with electricity, you will often be surprised with the brilliant results obtained in one case and the utter failure in another. Conditions that govern the prognosis in general should be considered in giving the probable effect of electricity. If electricity is to be beneficial, it will be manifest in the first two or three weeks of the treatment, and, if such improvement does not occur in that time, it will be useless to continue it longer; also, if, after improvement has continued for some time, it ceases, it will be useless to continue.

The time one should begin treatment varies with the severity and circumstances of the case. So long as the case is improving as fast as can be expected, we should not interfere with nature's process; but, if the improvement is slow or has entirely stopped, we should then try electricity. In cases of ordinary severity, three weeks should elapse before treatment is begun.

Galvanism should first be applied to the head in such a manner that the lesion is in the direct line of the current, so as to get its catalytic effect. The current should be allowed to run in various directions, thus increasing the catalytic effect; but the action of the positive pole should predominate on the side of the lesion, for that pole has a more energetic action on the cerebral substance. It is a question whether we should apply the electrodes so as to increase or decrease the blood supply to the head, for our knowledge of the exact condition present is so scanty that a rational conclusion can not be reached. From a purely theoretical standpoint we might conclude, if the person was young and plethoric, with signs of congestion of the head, that to decrease the blood supply would be advantageous. While, on the other hand, if the patient was old, poorly nourished, and anæmic, the electrodes should be so placed as to increase the blood supply. For the method of thus placing the electrodes, I refer the reader to cerebral anæmia and hyperæmia.

Central galvanization should always be given. The paralyzed muscles and nerves should be stimulated by carefully going over all the motor points and causing the muscles to contract. This may be done with either the galvanic or faradic currents. All the various symptoms and conditions should be treated according to their indications. If anæsthesia exists, it will generally be successfully treated by the faradic brush. If aphasia is present, it should be treated by placing the negative electrode of a galvanic battery on either side of the larnyx and the other over the third frontal convolution of the left side, which is the center of speech. If paralysis of the tongue exist, an electrode, connected with the negative pole of a galvanic battery, should be applied to the tongue and interrupted, so as to cause good contractions of it several times at each treatment.

The eyes are generally weakened by an attack of cerebral apoplexy, and will often be much improved by applying one electrode of a galvanic battery over the closed lids, the other electrode should be placed on the back of the neck, or, better still, over the site of the lesion, if this can be ascertained, and a current that is just perceptible allowed to pass for three or four minutes. The electrode on the lid of the eye should be alternately negative and positive. The same may also be said of the hearing, but here again a much weaker current will have to be given, or a sickening vertigo will be produced.

Chronic Bulbar Paralysis.—While electricity has as yet given at the best but poor results in bulbar paralysis, it has given better results than any other form of treatment, and should, therefore, be given a thorough trial. If the disease is in its incipiency, electricity may cut it short, or at least retard its progress; but, when far advanced, it will be of no use, as the manipulation around the mouth and throat necessary for a proper treatment may cause irritation and thus aggravate the whole condition.

The best method of treating it is to give transverse and longitudinal currents through the head, so that the medulla oblongata is in a direct line with it. The positive electrode should then be placed upon the occiput, while the negative pole, armed with an electrode of the required shape, should be placed on the lips, tongue and palate, and as strong a current as can be borne should be given and continued for three or four minutes. The negative pole is then attached to an ordinary electrode and held on either side of the throat externally, and a current strong enough to produce movements of deglutition passed for one minute on each side. This treatment should be given every day, or at least three times a week.

Disease of the Spinal Cord.—The principles which govern the application of electricity to the spinal cord have been given in Chapter II. Large electrodes should be used, unless we wish to increase the density in a certain spot, when an electrode of the required size and shape should be selected. When we desire to give the current lengthwise of the cord the electrodes should be large and placed some distance apart, for the further apart they are the deeper will the current penetrate (see Human Body as a Conductor). To affect the nutrition more directly we should place one large electrode on the abdomen and the other over the diseased part when it is confined to a single lesion, and labile over the whole spinal cord in general spinal disease.

Myelitis.—Although the prognosis in myelitis will depend largely upon the severity of the attack and the length of time elapsed since the beginning of the disease, it is generally good, and improvement varying from slight increase in motion to complete recovery is the rule and not the exception. The chronic or subacute stages only are amenable to electrical treatment. We should at first try to improve the nutrition of the cord and absorb any deposits that may have been thrown out. This will best be accomplished by placing one large flat electrode on the abdomen and with the other give both labile and stable applications over the diseased portion of the spine—both poles of the battery being applied to the spine at the same sitting, so as to increase the catalytic action of the current. To overcome the paralysis which exists, the positive electrode should be placed over the diseased portion of the spine, while with the negative interrupted applications should be made to all the motor points of the paralyzed limbs with sufficient strength to cause contractions of the muscles. If the whole section of the cord has been inflamed, anæsthesia will accompany the motor paralysis, and, if the lumbar portion of the spine is the seat of disease, vesical and rectal paralysis will be the result.

The anæsthesia will best be treated by the faradic brush, which is a metallic brush attached to one pole of a faradic battery, and, with a strong current, rubbed vigorously over all the parts that are so affected. This should be given in connection with the other treatment, as it undoubtedly exercises a beneficial influence over the cord by reflex action. In paralysis of the bladder the anode should be placed over the diseased portion of the spine and the cathode just above the pubes, a strong current passed, and several interruptions made; or one pole of a faradic battery may be introduced by means of an insulated electrode into the bladder and interruptions made (see Paralysis of the Bladder). The rectum should be treated in a similar manner. One large electrode (positive) connected with a galvanic battery, is placed over the lumbar region, and the other (an ordinary electrode) is pressed firmly against the sphincter, and several interruptions made.

Spinal Meningitis should receive the same treatment as myelitis. Some authors recommend an additional application with the faradic brush over the spine to produce counter-irritation.

Locomotor Ataxia.—Good results, as a rule, will follow electrical treatment in locomotor ataxia. It will, almost invariably, temporarily arrest the progress of the disease and relieve many of the symptoms accompanying it ; but these results are seldom permanent, lasting only for a few weeks or months. I believe a permanent arrest of the disease is sometimes achieved in those cases that are taken early, and, according to my experience, are not associated with syphilis, but are the result of injury. Erb advises applying one electrode in the sub-aural position, while the other is rubbed up and down the spine ; to this he adds labile application to the muscles of the lower extremities with the cathode—the anode resting on the sacrum. To relieve the pain in the limbs, he applies the anode over the affected nerve where it is given off from the spine, while the cathode is placed over the seat of the pain. He uses a current of medium strength, varying from fifteen to to thirty cells. Rumpf recommends daily faradization of all the muscles of the body. Notwithstanding the high authority from which this is taken, I must say I have seldom seen any benefit in locomotor ataxia by applying electricity to the muscles, and the faradic current very often irritates the patient, and thus aggravates the symptoms. The best method I have found is to give an ascending current along the spine. The negative electrode is placed on the back of the neck or in the sub-aural position, as recommended by Erb, and the positive stable on the sacrum for three or four minutes, and then moved slowly up and down the spine. A strength of current varying from ten to twenty milliamperes should be given every other day, the whole treatment lasting from five to eight minutes.

Lateral Sclerosis.—The treatment recommended by Erb in locomotor ataxia, and which is given above, I believe is the most successful application for lateral sclerosis of the spine.

Acute Anterior Poliomyelitis.—Good results, as a rule, will follow electrical treatment of acute anterior poliomyelitis. Even a long time after the disease has occurred and left some paralyzed muscles, if these muscles have retained their electro-excitability, they can be very much benefited by electricity. I saw a case of paralysis of the anterior tibial muscle of six years' standing, which was the result of an attack of acute anterior poliomyelitis, completely cured by placing the anode over the lumbar region and the negative over the motor points of the muscle and causing several contractions daily. The result will be better in adults than in children, but in all cases the treatment should be systematic from the first, and continued as long as there are any signs of improvement. The best method of treatment is to place the anode over the diseased portion of the spine—lower cervical; when the arm and lower dorsal and upper lumbar, when the legs are affected, and with the cathode go over the motor points and cause vigorous contractions of the diseased muscles. It is also well to rub the cathode over the integuments of the affected muscles until the skin shows signs of irritation.

Progressive Muscular Atrophy.—In this disease electricity may have only a slight effect in temporarily arresting the progress of the disease, but it has been known to effect a cure. We should first try to influence the irritation of the spine by applying both poles to it, thus increasing the nutritive effect of the current. Remack recommends applying the galvanic current to the spinal cord, especially to the cervical region. Next we should try to keep up the nutrition of the muscles. The galvano-faradic current may be substituted for the galvanic. The application should be made daily, causing very slight contractions. The motor points should be carefully gone over with the cathode, armed with a medium-sized electrode, while the anode is placed over that part of the spine which gives origin to the nerves which supply the affected muscles. Great care should be taken in giving this treatment, or harm may be done by fatiguing the remaining fibers of the diseased muscles. Only use a current of just sufficient strength to cause the slightest contractions, and do not prolong it more than half a minute to a muscle. It is a good idea to go over first one motor point and then another, and so on until all have been treated, and then return, thus giving the muscles a chance to rest. The faradic current may be used for this purpose, as the faradic excitability is not lost.

Spinal Irritation.—Electricity has been successfully used for many years in irritability of the spine. Great care should be taken when galvanism is used, or the treatment will be followed in a few hours by exhaustion, accompanied by profuse perspiration. Various methods for the application of the galvanic current have been successfully employed. One of the best is to apply the anode over all of the sensitive points, which should be diligently sought for, and the cathode placed on the abdomen. The current used should at first be very weak and of short duration, and then it should be gradually increased, if the patient bears it well, at each application.

My experience is that static electricity is by far the most effectual method of treating irritability of the spine. The direct spark should first be given over the back, avoiding the tender spots; but, after a few treatments, light sparks can be given over the tender spots as well. The spark should be of sufficient intensity to produce decided irritation of the integument of the whole back. This treatment can be given daily without any of the bad results which are liable to occur with galvanism. It will often happen that the back will not respond to a plaster placed on it for the purpose of producing irritation; but, after a few treatments with static electricity, it will respond very nicely. Static electro-massage or general faradization is often a beneficial adjunct in this treatment.

Functional and General Diseases and Diseases of Uncertain Origin—Neurasthenia.—Among the many remedies for this disease electricity is one of the most important. Occasionally brilliant results will be achieved by electrical treatment. As a rule, however, the improvement will be slow and steady, and rarely no results at all will be attained. In that form known as cerebral neurasthenia, galvanism should be applied to the head, and with particular reference to influencing the circulation, for this disease is generally associated with anæmia or hyperæmia of the brain, which should be treated according to the rules laid down on pages 91 and 92.

In spinal neurasthenia a strong descending current should be applied (anode on the back of the neck, cathode on the sacrum), followed by labile application to the spine with the cathode. In the general form of neurasthenia first give central galvanization and follow with general faradization, or, better still, by general electrolization, as described in General Therapeutics. Both central galvanization and general electrolization may be given during the same treatment, if the patient is strong enough to bear it without fatigue, but, if not, the applications should be made on alternate days until she gains the required strength.

If the patient is very weak, the treatment may be given in

bed ; but in all cases it is absolutely necessary that she lie down for from one to two hours following the application. The various symptoms that accompany this disease should be treated according to their indications. Constipation very often accompanies it, and should be treated as directed on page —.

Sick headache and cardiac symptoms should receive their appropriate treatment. Hypochondriasis very often accompanies neurasthenia, but it need not receive any special attention. In giving electricity in neurasthenia, as well as in hysteria, you should be very careful not to annoy the patient at first by too strong applications. Begin with weak currents and short applications, and gradually increase until the desired strength is obtained. The treatments should be given daily.

Hysteria.—This disease is often the object of electrical treatment, and meets with varying success. To give the methods that have been recommended for all the various conditions arising in hysteria would fill a volume. One general plan should be followed. Those conditions such as constipation, anæsthesia, irritation of the spine, and neuralgia, should be treated by the method recommended under their respective heads.

For the feigned diseases, paralysis, etc., treatment may be given as if they were real ; but the direct static spark is the best method of treating all such complications.

Static electro-massage is the best treatment for the contractures of hysteria ; but in the absence of the static battery, they may be treated the same as contractures in other diseases ; that is, pass a stable galvanic current through the contracted muscles and a strong faradic current through the antagonistic muscles. It is also good to give a general treatment daily with static electro-massage and at the same time draw a few sparks from the spine, epigastrium and limbs, or, if this is not convenient, give general faradization.

My experience has taught me to rely on the static battery in hysteria. Its success may be due to the mental effect produced by the brilliant spark ; but it makes no difference so long as the patient gets well, whether the cure is due to the mental or physical effect of the treatment. The current should be weak at first, and then gradually increased.

Never distress the patient by too strong an application, or you will destroy her confidence. No definite prognosis can be given. Sometimes brilliant results will be achieved when least expected, and at other times total failure.

Epilepsy.—In treating this disease we should at all times try to get the nutritive effect of the electric current on the brain by applying it obliquely and transversely through the head, and then longitudinally (the anode on the forehead and cathode on the occiput). As a good adjunct to this may be added central galvanization, with occasionally general electrolization. In order to achieve success in epilepsy by electrical treatment, it should be continued for a long time and given three times a week. I believe success will often follow a persevering systematic treatment. To insure a non-recurrence the treatment should be continued for at least six months following the last seizure.

Chorea.—Chorea is often the subject of electrical treatment. Its good effects are confined chiefly to recent cases in children, in which a cure can, as a rule, be expected ; but it is not generally successful in recent attacks in adults, or in chronic cases. The method of treating it is first to galvanize the brain by passing the current obliquely, transversely, and longtudinally through it, as in epilepsy ; next give labile applications over the cervical sympathetic with the anode, the cathode resting over the seventh cervical vertebræ. To this may be added a descending current to the spine (cathode on sacrum, anode on back of neck). The whole treatment should occupy from six to eight minutes and may be given daily, or at least three times a week.

Static electricity has been highly recommended for this disease ; but, in all the cases that the author has tried it, it seemed to aggravate rather than relieve the symptoms. The method used was to draw sparks from the spine and epigastrium.

Spasms.—Great contradictions exist in the reports of cases of different forms of spasms by different authors. Some claim to have had the most brilliant results, both palliative and curative ; while others claim that even palliative results are rare. The author's experience is that palliative results are often obtained, while cures are seldom made. The best results will be obtained by getting the anelectrotonic effect of the current on the affected nerves and muscles. A large electrode connected with the cathode of a galvanic battery is placed over the sternum. The anode should be first stable over the origin of the affected nerve, and then stable over the muscles (the electrode used should be large enough to nearly cover the muscle). In both cases the current should be slowly raised to its maximum, which should never be strong, and, after it has been continued from one to three minutes, as gradually decreased. In order to avoid the little interruptions that are liable to occur in turning the cells on and off, a water rheostat may be placed in the circuit and the current graduated by this means. If this treatment fails, other forms may be tried, such as passing the electric current through the diseased nerve to obtain its catalytic effect, or pass a descending current down the spine, both of which are recommended.

Voltaic alternatives through the affected nerves and muscles have been highly recommended in certain spasms, particularly facial. Excitation over adjoining healthy parts with the faradic brush is useful in cases of reflex spasm.

The direct static spark, both over the spasm or at some distant point, has been very highly recommended. In fact, I do not believe there is any form of administering galvanic, faradic or static electricity that has not been used, and if one fails, others should be tried.

Those spasmodic diseases in which electricity is most successful are trismus, facial spasms, blepharospasms and torticollis, and in these cases, when of recent origin and not depending on any organic lesion, a cure will occasionally be made. If the antagonistic muscles are weak, they may be treated with the interrupted galvanic current or by the faradic.

Contractures will be best treated by passing a stable galvanic current through the affected muscle, and treat the antagonistic muscles with the interrupted galvanic or faradic current. If the contracted muscles do not yield to the stable galvanic current, other methods may be tried, such as voltaic alternatives or vigorous faradization. The static battery has been of use in this disease, particularly when of hysterical origin (which see).

Tremors.—When tremors are due to toxic causes electricity may often be beneficial. The affected muscles should be vigorously treated with the galvanic, or, better, with the galvano-faradic current and by passing a galvanic current in various directions through the corresponding nerve centers. Paralysis agitans is the most severe form of tremors and is not amenable to electricity. The same is also true of tetanus.

Diabetes Mellitus and Insipidus.—In diabetes mellitus elec-. tricity has had but little effect, although a few successful cases have been reported from reliable sources. The treatment followed in the cases that were benefited differed somewhat, but all applied galvanism to the central nervous system. Central galvanization was mostly used and answers the purpose better than any other form of administration. With diabetes insipidus more has been achieved. The method of treatment that has, I believe, been the most successful is first to apply the galvanic current to the head so as to affect the medulla oblongata ; then place a large electrode connected with the cathode over the kidneys and give labile application with the anode over the cervical and dorsal spine ; now remove the anode to the epigastrium and give a strong stable current for two minutes. The whole treatment should last about eight minutes, and may be given daily or at least three or four times a week.

Migraine—Sick Headache.—In sick headache electricity will often cut short an attack, and, if the treatment is continued, will occasionally produce permanent benefit by lessening the frequency and duration of attacks, but rarely produces a cure. Various methods of application have been recommended, but the one that will usually be successful in cutting short an attack is to place a large electrode connected with the cathode over the epigastrium, with the anode stable over the casserian ganglion for from one to two minutes, and a current of medium strength passed. This may be done by applying the current to both of the ganglia at the same time by means of a bifurcated cord, or it may be first over one and then over the other. Next pass a current longitudinally through the brain (anode on forehead, cathode on occiput). If the eyes are the seat of pain, the anode may be applied over the closed lids. The treatment may be continued from five to ten minutes, when the pain and other symptoms will generally disappear. If this is not successful, others may be tried. The faradic current may be used instead of the galvanic, but it is generally inferior. The attack may be permanently cut short by the one treatment; but, as a rule, the symptoms will return in from five to twenty minutes, according to the severity of the attack, in which case another application should be given. If the treatment is successful the period of relief following the second application, if not permanent, will be much longer than before. Three or four such applications may be required before the attack is arrested. The same treatment may be given between the attacks to prevent their recurrence.

Exophthalmic Goitre.—This disease has been successfully treated by administering galvanism to the sympathetics. The method used is to place an electrode connected with the cathode over the cilio-spinal center above the seventh cervical vertebra; the anode should be placed over the auriculo-maxillary fossa, and, after holding it stable for a minute, labile application should be made all along the inner edge of the sterno-mastoid muscle, using a current of medium strength. The next step is to remove the anode to the position occupied by the cathode and place the cathode over the solar plexus: pass a very strong current for one or two minutes. This is Beard and Rockwell's method, and is the most successful used by the author. They recommend, when this treatment is not successful, rapidly increasing and decreasing the current strength, by means of a water rheostat, during the application. The puncture has also been recommended, and with good results; but, as the method used does not differ from that given for goitre, the reader is referred to page 136.

Writers' Cramps.—Under this head is included all the professional functional neuroses, whether they occur in writers, musicians, telegraphers, or artists. The electrical treatment should always be given in conjunction with massage and gymnastic exercises. The best method of treatment is to place the anode over the cervical enlargement and give labile applications over the brachial plexus and then over the affected muscles with the cathode. The current should be of medium strength and given daily. If the case is of recent origin good results may be expected ; but if of long standing, it is generally incurable.

Anæsthesia.—In treating anæsthesia, we should first try to get the catalytic effect of the galvanic current through the seat of the disease, if that can be made out, so as to influence the nutrition. This can be accomplished by passing a current through the diseased part in various directions. Next the parts affected with anæsthesia should be treated. This may be done by labile applications with the cathode of a galvanic

Fig. 53. current, or with the faradic current, using the metallic wire brush (Fig. 53) as the exciting electrode.

For this purpose the parts should be very dry and the current given strong enough to produce crackling on the surface of the integument. The direct spark, according to the author's experience, is the most successful method of overcoming anæsthesia. The treatment should be given daily.

Neuralgia.—The method of treating neuralgia is to first try to get the anelectrotonic effect of the galvanic current in the painful parts. The cathode is placed on the origin of the diseased nerve, the anode on the seat of pain, and the current raised to its maximum, which should not be very strong, and, after it has continued for two or three minutes, gradually decreased, always being careful to avoid any sudden interruption of the current. Painful points along the nerve should be sought for and treated with the anode in the same manner. In most cases of simple neuralgia the treatment will be very successful, but there are some cases which do better under the influence of the cathode, or when both poles are used alternately on the painful points, thus increasing the nutritive effect. In some cases, the interrupted galvanic current is more successful.

The faradic current should not be discarded in the treatment of neuralgia, for it will often cause relief when the galvanic fails. Just what the difference is in the cases that require the one or the other form of treatment, I am unable to say, but I believe those cases that require the anodic in-

fluence of the galvanic current to be those of simple hyperæsthesia, due, perhaps, to some slight molecular change in
the nerve; while those requiring the irritating effect of either
the cathode or the faradic current are caused by a more deep-
seated nutritive disturbance.

Rockwell has given an indication for selecting the battery to be
used. If pressure on the painful points aggravates the pain, the
galvanic current is indicated; if not, the faradic should be used.

Simple facial neuralgia is often amenable to electric treatment, but true tic-doloureux is not.

Cervico-occipital, cervico-brachial and intercostal neuralgia
should be treated in the manner stated above, and, when not
due to some organic changes, will as a rule be relieved.

Sciatica deserves special mention. The various forms of
treatment that have been recommended for sciatica are proof
that no one of them is often successful. The method recommended above (cathode on origin of the nerve and anode
stable over the painful part) should first be tried. If this is
not successful, two small electrodes may be applied six or
eight inches apart and the current gradually increased. The
electrode may be moved along the surface until the whole
nerve has in this manner been brought under the influence of
the current. One of the poles may be placed, by means
of a ball electrode, in the vagina or rectum in close
proximity to the nerve; but if the galvanic current is used,
the electrodes should be covered with sponge or chamois,
so as to prevent the electrolytic action causing an eschar on
the mucous membrane. If the part with which the rectal or vaginal electrode comes in contact is particularly sensitive, the electrode should be attached to the anode and the
cathode either placed over its origin or down on the thigh,
according to the judgment of the operator. If all of these
methods fail, the galvano-puncture may be tried. The anode
is usually applied by means of a large electrode over some
portion of the nerve. The cathode is attached to a fine
needle such as is used for removing hairs, and which is insulated within a few millimetres of the point with shellac. This
is introduced at several points along the nerve, and a weak
current passed for about one minute to each puncture.

In treating neuralgia, relief will follow for a certain length of time after the first application, then the pain will return, at which time another application should be given ; and, if the treatment is to be successful, the period of relief following each application will be longer than the previous one. In giving the prognosis of electricity in neuralgia, one thing should be taken in consideration, and that is the depth of the affected nerve. When a person is very fleshy, or the nerve is deep-seated, electricity will not be as beneficial as when the nerve is superficial, as the current will be distributed in the overlying tissue and comparatively little of it will reach the affected nerve.

Diseases of the Peripheral Nerves—Paralysis of the Facial Nerve.—This affection may be either of central or peripheral origin. When due to hæmorrhage of the brain, it requires the same kind of treatment as cerebral hæmorrhage (which see). When it is accompanied by some other disease, such as chronic bulbar paralysis, it should be treated as a symptom of that disease. When of peripheral origin this paralysis is generally produced by a blast of cold air on the side of the face, causing effusion into the sheath, and consequently compression of the nerve fibers, although it may be due to injury or disease at the base of the skull, along and through which the nerve passes, or to disease of the nerve itself. It is when due to the former cause that electricity is so efficacious.

The method of treatment is to place the anode behind and just below the ear at the point of exit of the nerve, while the cathode is first given labile over the face, and then each motor point is carefully gone over with a small nerve electrode and the current interrupted several times. The current should be strong enough to cause muscular contractions, which should be slight in the first stage of the disease, but stronger as the muscles increase in strength.

As has been stated in Electro-Diagnosis, the electric test will give a clue to the prognosis. If no sign of R. D. is present, the patient will recover in from two to four weeks ; if it is slightly present at the end of the second week, it will require from five to eight weeks, and, when pronounced, several months will pass before recovery is complete. If the disease has been of long standing there may be permanent impairment.

Diseases of the peripheral nerves in various parts of the body, due either to cold, traumatism, or disease, will require about the same treatment as that given for the seventh nerve. First we should pass a galvanic current in various directions through the diseased parts to cause absorption, if there be any deposits or effusions, and to improve the nutrition. In case of neuritis, when the nerve is sensitive and painful, the best results will at first be attained by applying the anode stable over that part of the spinal cord which gives origin to the affected nerve, and the current gradually increased to as strong as can be borne, and then as gradually decreased. Second, we should stimulate the muscles and nerves by strong labile applications with the galvanic current.

Paralysis of the Oculo-motor, Trochlear and Abducens.—These nerves are frequently the seat of paralysis, sometimes centrally, but generally of peripheral origin, in which case they are often successfully treated by electricity. The anode should be placed on the back of the neck, and a small nerve electrode attached to the cathode should be placed over the insertion of the muscle supplied by the diseased nerve over the closed lid. A current first stable for one or two minutes is passed and then interrupted a few times. The current should be as strong as can be comfortably borne.

Laryngeal Nerve.—This can be treated externally by placing the electrodes on the neck each side the trachea, or internally by applying either one or both directly to the paralyzed part. I do not believe there is anything gained by the latter method, as the current used is necessarily much weaker, it is difficult to apply, and is certainly annoying to the patient. It should not be tried until the external method has failed. The method of administering it is to place the anode (if the galvanic current is used) on the neck, and with the other give labile and interrupted application over both sides of the larynx. Some recommend the use of galvanism exclusively, but I have known bad symptoms, such as profuse perspiration, syncope, etc., to follow its application. I would, therefore, advise it to be used cautiously at first, and, if these symptoms occur, try the faradic. The galvano-faradic is more effective than either one separately. I have seen good results from the static spark when applied over the larynx.

When the current is applied internally the faradic battery is always used. Place one electrode on the back of the neck and touch the diseased muscles with the laryngeal electrode, which should be connected with the other pole of the battery. Some use a double laryngeal electrode, by which both poles are placed side by side in the larynx. In either case the current should be strong enough to produce contractions.

Hysterical aphonia may be treated in this manner, but the current should be strong enough to produce an unexpected cry.

Toxic Paralysis—Diphtheritic Paralysis.—When this disease attacks the pharynx, it requires the same treatment as bulbar paralysis and paralysis of the pharyngeal nerve, which are given in another part of this chapter, and to which the reader is referred. When it attacks other parts of the body it should be treated on the general principles laid down for peripheral paralysis, and that is, to pass the galvanic current through the seat of the disease, if this can be made out, and give labile and interrupted application over the paralyzed parts.

Lead Paralysis.—Here again we have the electrical test giving a clue to the prognosis. R. D. is nearly always present to a certain degree in more or less of the affected muscles. It may not be present in all, and in those recovery will take place in a few weeks, but is seldom complete, and a relapse is liable to occur. The method of treatment is to place one large, flat electrode over the lower cervical and upper dorsal region, and another large electrode on the sternum. These should be alternately positive and negative, and a strong current for three or four minutes is passed, but the current should be reduced before the polarity is changed, so as not to shock the patient too severely. The anode is then left over the back, and with the cathode interruptions of the current are given over all the motor points of the affected muscles and nerves, and vigorous contractions produced. Sometimes it is necessary to apply both electrodes to the paralyzed muscles and nerves to cause contraction.

Diseases of the Eye—Disease of the Lids and Conjunctiva.—
Electricity has been very successfully employed in blepharitis,
trachoma and follicular conjunctivitis. The best method of
treating these diseases is by the bipolar method, given by
means of the double electrode, which is perfectly insulated,
except at the tips, where two bulbs, four and a half millimetres in diameter, are fastened one-half of an inch apart (Fig.
54). This electrode is curved at about the same angle as a
urethral bougie, and with it all parts of the conjunctiva can be
easily reached.

In a case of blepharitis, if the diseased part is covered with
scales, they should be removed before beginning the treatment. The bipolar electrode is then rubbed over the surface.
In a case of trachoma, the electrode should be rubbed over all
the surface of the conjunctiva that is diseased. The treatment
should be given as strong as can be comfortably borne and

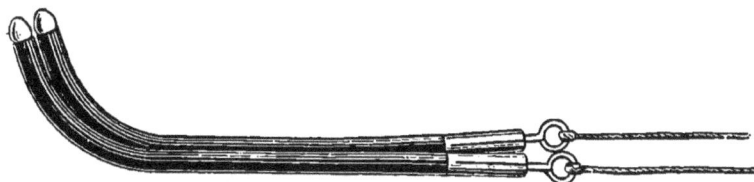

Fig. 54.

continued for four or five minutes, and not oftener than twice
a week, and in some cases once a week is better still. In follicular conjunctivitis, when the follicles are very large, they
will have to be punctured with the negative pole. The small
needle used for the removal of superfluous hairs should be
used for this purpose, and each large follicle destroyed. Cocaine may be put in the eye, thus making the operation pain-
less. The number of treatments required will be from ten to
twenty, according to the severity of the case. In diseases of
the *cornea, retina,* and *optic nerve,* we depend entirely upon
the nutritive effect of the galvanic current. This will be best
obtained by placing a small, flat electrode, covered with sponge
or chamois, over the closed lid, and the other, a medium-sized
electrode, on the occiput. The anterior electrode should be
first connected with the negative, then with the positive, and,
last, again with the negative. The current should be gradu-

ally raised to its maximum, after the electrodes are placed in position, and gradually reduced before the change in polarity is made, after which it is as gradually raised and lowered again, and so on. The treatment should continue about ten minutes, letting the negative rest on the eyelid about twice as long as the positive. The disease for which this treatment will be found most beneficial is asthenopia, which it often relieves in a very few sittings. In some cases, where the galvanic current has failed, the faradic current, given in the same manner, has been successful. The galvanic current given in this manner is also useful in the sequela of inflammation of the optic tract, and it is very successful in that form of opacity of the vitreous body which is due to infiltration from any cause or extravasation of blood ; but is of no use when it is the result of a degenerative process. Primary atrophy of the optic nerve or retina, or cataract, according to my experience, cannot be permanently benefited by electricity. The eye is merely stimulated, and the patient almost invariably thinks he sees better ; but, on testing vision, it is discovered that there is no improvement.

Paralysis or spasms of the orbicularis palpebrarum, oculomotor, trochlear and abducens should be treated according to the principles laid down in another part of this chapter for those diseases.

Diseases of the Ear.—Numerous attempts have been made for several years to cure diseases of the auditory apparatus by electricity, but very little has as yet been accomplished. This is undoubtedly due to the small amount of current that can, under the most favorable circumstances, reach the inner ear. As stated in Electro-Physiology, the greater the angle of the current to the sagittal suture the more readily is vertigo produced ; consequently, in treating the ear, a light current has to be given, and when we consider that the auditory nerve is in a bony cavity, thus making the resistance very great, it will be understood that the amount of current that reaches the nerve is small.

There is, however, one disease which deserves special mention. This is tinnitus aurium, when of nervous origin. It is a very annoying condition, and, as no other method of treatment is successful, it is all the more important in this connection. Place the ear electrode (Fig. 55) in the ear, after the external auditory canal has been filled with warm water. The other electrode may either be placed on the tongue by means of the tongue electrode, or on the opposite side of the head. From all theoretical considerations it would seem that the positive pole should be attached to the ear electrode, and the current gradually increased to its maximum and as gradually decreased, as it is considered an hyperæsthesia of the auditory nerve. While this relieves more cases than any other method, it is found that often the negative pole acts the best, and sometimes an opening or closing acts better than the stable application. The only way is to use the method that affords the most relief. After one treatment relief will follow for a certain length of time, when the tinnitus will return, but is again relieved by the current, and this relief con-

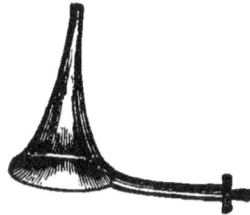

Fig. 55.

tinues for a longer time than before, and so on until the disease is cured. If the first period of relief continues for some time, and, when it does return, is much less marked, recovery will take place in a week or so; but if the period of relief is short and the symptoms return with all their vigor, months will elapse before recovery takes place, or the case may be found to be incurable.

There is generally some deafness accompanying tinnitus aurium, and this is relieved as the symptoms of the latter disappear.

Nervous deafness may be relieved by the same method of treatment.

Diseases of the Heart and Lungs—Asthma.—Different authors recommend different forms of application of electricity for the relief of asthma, some the galvanic and others the faradic, but nearly all agree that the pneumogastric nerve must in some way be stimulated to cause relief. In order to relieve an acute attack, place an electrode on each side of the neck below the angle of the jaw, and just in front of the sterno-cleido mastoid muscle, and pass a strong current from a faradic battery for thirty minutes twice a day. If this is not successful, the positive pole of a galvanic battery should be applied to the occiput, and with the negative give labile applications from the sub-aural region down to the sternum. Sometimes good results are achieved by reversing the poles. This I believe is also the best treatment to be given between the attacks to prevent a recurrence and should be applied daily. De Watteville recommends the galvano-faradic current to be given in this manner; but, according to my experience, the good effect of the faradic current is limited entirely to the relief of the attack at the time. It has been known to do much harm when given between the attacks.

Angina Pectoris.—Duchenne attained good results from cutaneous faradization with the brush electrode over the præcordial region and especially the nipple. By this method he succeeded in cutting short the attacks, and, by continuing the treatment, cured some of the cases. The most successful method of treatment, however, is with the galvanic current. Place a medium-sized electrode connected with the anode over the præcordial region, go over the sympathetics and down the spine with the cathode, as in central galvanization, and, at the end of the treatment, let the cathode rest stable over the last cervical vertebra for two or three minutes. I once saw the most brilliant results from this method of treatment. The current should be very weak at first and the sittings of short duration, but they may be increased with each treatment. The sittings may be given every two hours during the attack, and once a day for some time following it.

Nervous palpitation of the heart may be treated the same as angina pectoris.

Artificial Respiration.—This cannot be considered a disease of the heart or lungs, but can properly be considered in this connection. The galvano-faradic current is the best to use; but in the absence of the galvanic, the faradic may be used alone. Place one medium-sized electrode over the motor point of the phrenic nerve. This may be applied to only one side, but it is better to apply an electrode over both phrenic nerves by means of a bifurcated cord ; the other electrode, which should be of a large size, is placed over the region of the ensiform cartilage and a very strong current passed. This should be interrupted from three to five times a minute. If life has not been extinct for more than three minutes good results can be expected. The cases in which it is indicated are opium and carbonic acid gas poisoning or asphyxia from chloroform. In all cases electricity should be combined with the other methods of artificial respiration.

Diseases of Muscles and Joints—Muscular Rheumatism.— This disease is very successfully treated by electricity. The faradic battery may be employed by applying a current direct to the seat of pain. If the muscles are very painful, the current should be weak at first and gradually increased at each treatment until the required contractions are produced. The treatment may be given two or three times a day. Galvano-faradization given in the same manner is still more effective. If the galvanic current is used, the cathode should be applied to the seat of pain. Give labile application strong enough to produce contractions.

According to my experience, static electricity is by far the most successful in treating muscular rheumatism. If the pain is great, the indirect spark may be given at first, but the direct spark should be applied as soon as it can be borne. Sometimes one application of the direct spark may cure the patient.

Muscular Atrophy.—When this disease is due to non-use or pressure, such as follow fractures, etc., success can be expected to follow electrical treatment. Either the galvanic or faradic current may be used. The motor points should be gone over and contractions caused. This treatment may be given daily or three times a week.

Sprains, bruises and fractures will be greatly helped by electricity. I know of nothing that will relieve the pain and soreness of a sprain equal to electricity. I have seen patients suffering from sprains come in my office on crutches and after treatment walk out without a limp. The galvanic or faradic current may be used by passing a current directly through the affected part ; but, as with muscular rheumatism, the direct spark of the static machine is most successful. Electricity should never be given in sprains until the inflammatory stage has passed. In case of a fracture near a joint, that joint will often be very much stiffened and sore, which is due chiefly to a deposit around the fracture. This will be relieved by passing a galvanic current two or three times a week through the joint, the current being occasionally reversed to increase the catalytic action.

Articular Rheumatism.—No benefit accrues from treating this disease with electricity in its acute stage ; but marked improvement follows its use in the chronic stage. The galvanic current should be used. A strong current is at first passed in various directions through the joint, then the positive electrode is placed over the spine where the nerves are given off which supply the affected part, and, with the cathode begin below the joint and give labile applications upward. Sometimes there will remain a soreness after the joint has apparently returned to its normal condition, and which can not be removed either by the galvanic or faradic currents, but which will disappear after a few treatments by the direct spark of the static machine.

Rheumatoid arthritis is seldom benefited by electricity. It may, however, be palliated by the current. It should be treated the same as articular rheumatism.

Diseases of the Abdominal Viscera.—In treating diseases of the abdominal viscera we may wish to get the nutritive effect of the galvanic current, but we will have to depend chiefly on the contractile powers of the current upon the involuntary muscular fibers. For this the faradic current will be of most use.

If the primary coil furnishes a current of sufficient strength, it should be used. Still better .is the galvano-faradic current. The positive pole of the galvanic battery is connected with the negative of the secondary coil. Here we get the stimulating effect of the faradic combined with the deep, penetrating, continuous current.

Diseases of the Stomach.—*Gastrectasia*, or dilatation of the stomach, is sometimes successfully treated by passing a strong faradic current through it. A medium-sized electrode should be placed over each end of the stomach. Also one electrode of large size may be placed directly over the dilated part of the stomach and the other on the back. Currents should be given as strong as can be borne and continued for from fifteen to thirty minutes, and repeated daily, or at least three times a week.

Indigestion.—The same treatment may be successfully used in certain nervous and atonic forms of indigestion, particularly when accompanied by constipation.

Diseases of the Liver.—In functional disturbances of the liver I know of no remedy equal to galvanism. A medium-sized electrode (positive) should be pressed down under the ribs as near the liver as possible, and a large negative electrode on the back in such a position as to bring the liver between the electrodes; a current as strong as can be borne should be passed for ten minutes at each treatment, and repeated two or three times a week. It is sometimes astonishing how the symptoms, such as mental depression and uneasiness of the right hypochondria dependent upon it, disappear. After an acute hepatitis which has left inflammatory deposits, this same treatment is very efficacious. In such cases the current can be reversed with advantage. In the organic forms of liver disease I have never known any permanent good to be attained, but sometimes the palliative effect pays for the trouble of administering electricity.

Disease of the Spleen.—Electricity has long been successfully used in treating enlargements of the spleen. The principle of the treatment is simply to cause contraction of the muscular fibers of which the organ is composed. One large electrode is placed over the left hypochondrium and the other upon the back, so that the spleen lies between them, and a galvano-faradic or faradic current is passed. Very strong currents are required, and should be continued for one half hour daily. Electrolysis has of late been employed for enlargement of the spleen. The rules given under electro-abdominal puncture apply here. Treatments should be given about once in two weeks.

The author has never performed this operation.

Diseases of the Intestines—Constipation.—Undoubtedly the most common complaint of the human family is constipation. This is sometimes as difficult to cure as it is common, but the author has yet to get a single case which he has had a fair chance at, but what he has succeeded in curing with electricity. The galvano-faradic current is best; but the faradic will answer. Apply one electrode at the sphincter ani and with the other go over all the abdomen, using the most powerful current that it is possible for the patient to bear, and continue until the peristaltic movements of the intestines can be distinctly heard when placing the ear near the abdomen. In severe cases treatment should be given every day; in ordinary cases two or three times a week will suffice. Some physicians recommend the placing of one electrode in the rectum, but this is annoying to the patient, and I have never found it necessary, except in those cases where the principal fault is in the rectum, and this is usually caused by rectal injections.

In some very bad cases it may require some assistance to get the bowels to move at first, as electricity never acts as a cathartic. A glass of hot water before meals may answer the purpose, or if not, a few drops of the tincture of hydrastis may be added; but just as soon as movements begin, all such adjuncts should be gradually discontinued. These are necessary only in those desperate cases which have required an enema or a cathartic to produce a movement, and then only as an adjunct for the start, and should never be continued. I always require strict regularity with my patients, and sometimes regulate their diet.

In some cases of obstruction and intussusception electricity is of use. Several cases of both these diseases have been reported cured by good authority. The treatment is the same as for ordinary constipation, only more vigorous.

Prolapsus Ani and Paralysis of Sphincter Ani.—When of central origin, this disease has already been considered as a disease of the spinal cord. When it is local and caused generally by hæmorrhoids or constipation it is successfully treated by electricity. A large electrode attached to one pole of a faradic battery is placed over the sacral nerves, and the other, a rectal electrode, is placed just inside the sphincter and a strong current passed for ten minutes and given from three to six times a week.

Diseases of the Nose.—Hypertrophic and atrophic nasal catarrh has been successfully treated with electricity. The principle underlying the treatment is to get the galvano-caustic action on the hypertrophic forms and the nutritive effects in the atrophic forms. In the hypertrophic form a flat electrode (Fig. 56) is introduced, so that the blade comes flat-wise against the hypertrophied tissue, and the other electrode should be placed on some other part of the body; a current of from ten to twenty milliamperes should be passed for from eight to ten minutes. These treatments may be given once in two weeks, and about eight to ten treatments will be required to produce a cure. When the parts are hæmorrhagic, or when there is a profuse discharge from the hypertrophic parts, the positive pole of the battery should be attached to the electrode inside the nose; but when these complications do not exist,

Fig. 56.

the negative should be used. When the positive is used the instrument should be made of platinum. This treatment is inferior to the galvano-cautery, which is given elsewhere. In atrophic forms the treatment should be the same, except the current should not be so strong or the electrode may be covered with chamois. This treatment is always good in relieving the discharge and odor of ozena. Tumors of the nose or pharynx may be treated by electrolysis the same as tumors elsewhere. The negative electrode is introduced into the tumor, while the positive is placed on some indifferent part of the body. There is no pain or soreness following the operation. The tumor gradually shrinks up and disappears altogether. The same treatment should be applied to the tonsils, but the needle should not be introduced more than one-fourth of an inch.

Diseases of the Male Genito-Urinary Organs.—The diseases for which electricity is useful in this department are sper-matorrhœa, seminal emissions, impotence, incontinence of urine, paralysis of the bladder, gleet, and stricture of the urethra.

Spermatorrhœa and Seminal Emissions.—As these two forms of disease generally require the same method of treatment, they are included together. There might also be included in this section the premature discharge of semen in sexual intercourse when it is due to a nervous condition. I first give the patient either general faradization or central galvanization, whichever is found to be the more effective. Sometimes when one has previously produced good results and has ceased, a change to the other gives a new impetus to the case. Next introduce the largest sized Neuman sound (see Fig. 57) that the urethra will take, so that the bulb comes in contact with the opening of the vesical ducts. The other electrode is placed on the perinæum and a strong faradic current allowed to pass for from five to ten minutes. I do not believe in using the galvanic current for this purpose, as it is dangerous and is not so beneficial as the faradic.

Impotence.—The author has tried various methods of treating this disease with electricity, and when one method fails, others should be tried. He believes the most beneficial method is to use the galvano-faradic current. Place the electrode that is connected with the negative pole of a galvanic battery over the sacral region, and, with a good-sized electrode, press the testicles and penis firmly up against the abdomen, thus bringing the electrodes in firm contact with the under surface of the testicles and penis. This will often cause an erection during the treatment. If there is any bodily weakness, central galvanization or general faradization may be given the same as in spermatorrhœa. If the testicles are cold, relaxed and flaccid, with a diminished amount of secretion of semen, the most benefit will be attained by passing a galvanic current directly through them. If there is anæsthesia of the penis, which is quite often the case, it should be treated with the faradic brush the same as local anæsthesia of any other place. Treatments should be continued for from ten to fifteen minutes and repeated about three times a week. This is decidedly the most effective method I have ever employed, and I never knew a case to be relieved by any other form of electrical treatment when this has failed.

Vesical Paralysis.—Incontinence of Urine.—There are two methods of treating paralysis of the bladder and incontinence of urine, the external and internal. The external method consists of passing an interrupted galvanic or faradic current through the bladder by placing one electrode over the pubes and the other over the sacrum in paralysis of the bladder; but in incontinence of urine the sacral electrode should be placed on the perinæum. This treatment should be given to children, but in adults a much better method is to combine both external and internal. In paralysis of the bladder the internal electrode should be introduced into the bladder when it is partially filled with urine, and the other placed on the pubes or sacrum, and a strong current passed from a faradic battery. The external treatment might be given, using the galvanic current at the same sitting. In incontinence of urine the internal electrode should be passed in until it just reaches the sphincter muscle and a faradic current given. For this purpose I use the Neuman electrode (see Fig. 57), and use as large a ball as can be passed so as to come in contact with the muscle. This treatment may be given three times a week, and as strong a current as can be borne should be allowed to pass for from ten to fifteen minutes at each application.

Stricture of the Urethra.—Dr. Neuman has made an exhaustive study of this subject, and it is to him that we owe all of our present knowledge. His electrodes are made of steel wire covered with hard rubber and have a metallic tip which varies in size, running the whole American scale.

The method of operating is first to diagnose the exact size and location of the stricture; then choose a bougie from one to two sizes larger than the stricture, and place a rubber ring around it at such a distance that when it comes in contact with the meatus the bulb comes in contact with the stricture. Only the slightest 'pressure should be made against the stricture.

The negative pole should be attached to the internal electrode and the positive either applied to the groin or held in the hand of the patient; eight to ten cells are used and the

treatment continued for from five to fifteen minutes, when the
electrode will slip through the stricture. This treatment
should not be given oftener than once a week or once in two
weeks. The kinds of stricture this treatment is most successful in are those located just in front of the prostate and not of
very small calibre ; but it is not successful in those located
near the meatus or for an irritable or spasmodic stricture.
The latter varieties may be relieved by being treated in the
same way with a faradic current.

Diseases of the Skin.—Eczema and herpes have been successfully treated by electricity. The method used is to place
one pole of the galvanic battery (preferably the cathode) over

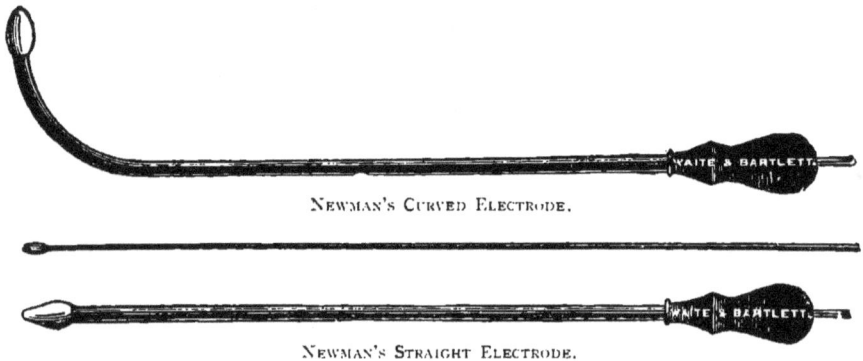

NEWMAN'S CURVED ELECTRODE.

NEWMAN'S STRAIGHT ELECTRODE.

Fig. 57.

the origin of the nerve that supplies the parts affected, while
the other is applied over the diseased part. This is best done
by means of a wet towel or piece of linen, which is folded so as
to cover the parts to be treated and the electrodes placed over
it. A strong current is allowed to pass for five minutes. The
above treatment is based on the supposed nervous origin of
the disease. Another method is to apply both poles to the
diseased parts by means of the double roller electrode (Fig. 58).
The best results will be attained when the galvano-faradic
current is used. Treatments should be given three times a
week.

Pemphigus may be successfully treated by the same
method.

Alopecia Areola may be successfully treated by applying

both poles of the galvanic battery, by means of the double-roller electrode, to the bald spot. The current should be quite strong and given for from five to eight minutes two or three times a week. Sometimes the growth of hair that follows this application is very rapid.

The loss of hair from fever or specific disorders may be treated in the same way.

Obstetrics.—Electricity, in the last few years, has at times been recommended for nearly every condition that arises during the maternal and puerperal state. The conditions for which it is found most useful are uterine inertia, deficient lactation and extra-uterine pregnancy.

Uterine Inertia.—In treating uterine inertia we depend upon the contractile effect of the faradic current on the muscular

Fig. 58.

fibers of the uterus. The best method of attaining this result is to place one rather large electrode over the abdomen, directly over the gravid uterus, and the other over the lumbar region. A strong current is then passed for from five to eight minutes, and, after an intermission, repeated. The length of the intermissions varies with the stage of the labor. If it be in the first stage, they should not be oftener than every half hour, while, if it be in the second stage, it should be given oftener. This possesses some advantages over ergot. The patient suffers no ill effects from electricity, and the os dilates very rapidly under the contractions produced by it, while just the contrary is the case with ergot. But an electric battery is cumbersome, and is generally out of reach when the physician wants it. As compared with the use of the forceps, when the condition of affairs justifies their use and the forceps can readily be applied, I would give the preference to the forceps, when they are in skillful hands; but if in unskilled hands, I would by far give preference to electricity. Postpartum hæmorrhage, due to a non-contracting uterus, may be treated in a similar manner. In a desperate case the electrode may be introduced into the vagina, or even into the uterus itself.

Deficient Lactation.—When this is due to deficient development, the treatment should be given before pregnancy, or not at all; but when due to other causes the author has succeeded in benefiting about 70 per cent. of the cases treated. If the patient is weak and anæmic, general faradization may be given in conjunction with the local treatment, which consists in passing a strong faradic current through the breasts for from ten to fifteen minutes. The number of treatments required varies from two to twenty, and should be given daily.

Extra-uterine Pregnancy.—The use of electricity in extra-uterine pregnancy has become familiar to all physicians. It, therefore, only requires a description of the method of administration here. The method most generally adopted is to use the interrupted galvanic current ; the negative pole, armed with a suitable electrode covered with sponge or chamois, should be introduced into the rectum or vagina (the one being selected that will bring the current in most direct contact with the fœtal nest), and the other placed on the abdomen over the tumor. A current as strong as can possibly be borne, with frequent interruptions, should be passed for from five to ten minutes. All that is required is to cause the death of the fœtus, and when this is accomplished treatment should be suspended. This may require but one treatment in some cases, while in others from five to ten will have to be given. They should be given daily, when possible; if this can not be done, apply as often as the patient can bear it. The diagnosis of the death of the fœtus requires great judgment, and, if no bad symptoms occur, the treatment had better be continued, so as to be on the safe side. There is one condition that I believe invariably follows the death of the fœtus, and that is the softening up of the mass. If the index finger can be run over the surface of the tumor while the fœtus is alive, the mass will be found to be very hard and the surrounding envelopes will appear to be on the stretch from the internal pressure. But in about forty-eight hours after the death of the fœtus this sack will become softened, as if some of the contents of the fœtal sack had become absorbed. Some physicians prefer the faradic current to the galvanic, but it will require more treatments with the faradic, it is more uncertain in its action, and it possesses none of the catalytic action of the galvanic current, all of which are very important in this condition ; and, on the other hand, it possesses no corresponding advantage. I, therefore, advise the use of the galvanic at all times. Just how the dead fœtus is gotten rid of by nature is not known, but it is supposed that it is first changed to a liquid state and then disappears by absorption.

Diseases of Women.—Electricity has a very important place in the diseases of women. In fact, it has made more rapid strides in this department in the last five years than in any other. The diseases for which it is used are both organic and functional. It relieves the symptoms in the functional and both the symptoms and lesions in the organic, but the relief of the former are always more marked than the latter. Electricity may be administered externally or internally. Formerly, when an external application was made, one of the electrodes was placed over the back and the other over the abdomen. When one considers that the female organs of generation are located below the brim of the pelvis, it is evident that the uterus only gets a small per cent. of the current thus passed. I have devised another method of placing the electrodes. One large electrode is so made that it fits the perinæum and curves upon the end of the spine (as in Fig. 59), while the other, an oblong electrode, is pressed well down over the pubes, the uterus thus being directly between the electrodes, and that organ, consequently, gets the greater part of the current. One or both poles may be used internally. Where one pole is used internally it may be applied to the cervix, in the vaginal cul-de-sac, or in the uterus itself, while the other is usually placed over the abdomen, but may, in certain cases, be applied to the back. When the galvanic current is applied and the internal electrode is placed in the vagina, the metal part of the electrode should be covered with chamois to prevent it from causing an eschar, but this is unnecessary when the electrodes are placed inside the uterus. When both poles are used internally a double electrode should be used. (Fig. 60 represents Apostoli's double electrode for vaginal applications, and Fig. 61 represents the same author's double uterine electrode.)

The diseases for which electricity is useful are amenorrhœa, dysmenorrhœa, menorrhagia and metrorrhagia, cellulitis, endometritis, metritis, subinvolution, superinvolution, ovarian irritation and chronic ovaritis, fibroids and ovarian tumors,

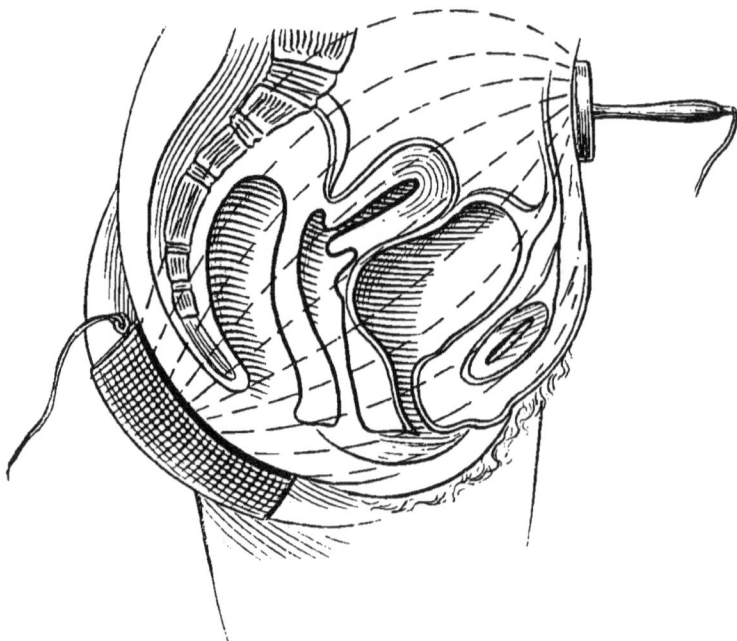

Fig. 59.

deficient development, stenosis of uterine canal, flexions and uterine displacements.

Amenorrhœa is generally relieved by electric treatment, but may prove very obstinate. In maidens the external faradic treatment should at first be tried. If this is not successful, the internal method may be used. I have never seen good results in this condition from applying the galvanic current externally, but the best of results may be expected by the internal use of galvanism. The negative pole should be introduced in the uterine canal and the anode placed over the abdomen; or the faradic current may be used internally in the same way. This is generally more successful than using double electrodes. If the internal electrode cannot be introduced, one may be applied to the cervix. If the patient's general health is below par, good results may be expected to follow general faradization, in connection with the local treatment.

Those who possess a static battery will get most excellent results from its use.

The direct static spark should be given over the pubes and lower part of the trunk. This treatment is particularly applicable to maidens and other ladies who object to internal treatment, for the static spark is always more successful than the faradic application externally.

Fig. 60.

Fig. 61.

Dysmenorrhœa.—In studying the reports of cases of dysmenorrhœa, one can not but be surprised with the variety of methods recommended and the varying results attained. This is due to the numerous and uncertain causes and the various classifications of the disease. It is seldom that two physicians diagnose a case of dysmenorrhœa the same. It is, therefore, impossible to lay down any one special method of treatment. If the patient is anæmic and poorly nourished, general faradization may be given for its tonic effect. If stricture of the os is the cause, it should be treated according to the principles laid down elsewhere for that condition; but in the majority of cases, the best results will be attained by putting the negative electrode of the galvanic current into the uterus and the positive over the abdomen, and a current passed of from twenty to fifty milliamperes in strength, and if of the membranous type from 150 to 200 milliamperes should be given for from eight to ten minutes three times a week between the menses. This will generally require an anæsthetic, as the endometrium is very sensitive in membranous dysmenorrhœa. In cases that are purely neuralgic, the positive pole should be attached to the intra-uterine electrode, which should in this case be made of platinum. If the internal treatment cannot be given, the external may be the same as in amenorrhœa. My experience has been that the galvano-faradic current is the most effective, but either of the others may be given separately.

Non-development of the Female Sexual Organs.—If the uterus and ovaries are congenitally so deficient that they are not recognizable, it will be useless to undertake any kind of treatment tending to their development ; but, when they are developed to a certain extent and lack the required stimulus to cause regular ovulation and menstruation, they can generally be brought up to the required standard by the continuous and judicious use of electricity. The prognosis should be particularly good, if there are symptoms indicating the menstrual period at stated intervals. It will be best to begin the treatment by external applications, one electrode on the perinæum and the other over the abdomen. As previously described, the abdominal electrode should be first pressed over the uterus, and then over each ovary. The galvano-faradic current will be found to be most efficacious, but the faradic may be used alone. If this is not successful, a ball electrode may be placed in the vaginal cul-de-sac, and the other over the uterus and ovaries, the same as with the external method, and the faradic current used ; or, if possible, one electrode should be introduced into the uterus. The galvano-faradic current can be used to advantage ; in which case the cord connected with the negative pole of a galvanic battery should be attached to the uterine electrode. As strong a current as can be borne should be given for from ten to thirty minutes three times a week. The time required for a cure varies with the amount of stimulus needed, and may vary from one to six months.

Superinvolution requires the same treatment as the non-developed uterus, with perhaps the occasional addition of general faradization, if the patient is in poor health.

Sub-Involution, Chronic Metritis and Endometritis.—The method of treating sub-involution by electricity is both simple and effective. A double electrode (Fig. 61) is placed inside the uterus, so that a current passes from one end of the organ to the other. A strong current from the secondary coil, which is composed of short, thick wire, should be allowed to pass for from five to ten minutes, when the uterus will be found contracted and the cavity much smaller than before the treatment. Of course all this contraction does not remain ; but abundant experience proves that part of it remains, and, by giving a treatment two or three times a week, in a few weeks a large sub-involuted uterus will be reduced to its normal size. If the disease is of long standing and the uterus has become hard and tender, it should be considered as chronic metritis and treated according to the method given below for that disease.

In chronic metritis and endometritis we will have to depend entirely on the chemical and catalytic effect ; but with this, as well as in other diseases, indications are given for the pole to be used inside the uterus. The method of treatment is similar to that in fibroid tumors—the active electrode of platinum being introduced into the uterine canal, while the other pole is attached to a large electrode placed over the abdomen. It will be remembered that the positive pole produces a dry, white, contractile eschar, which is hardening, drying, and markedly hæmostatic in its effect. It should, therefore, be used as the internal electrode in those forms of metritis and endometritis which are accompanied with hæmorrhage, also in the ulcerated and granular form, and it is the most successful method of treating uterine leucorrhœa that the author has ever used. The negative pole produces a soft eschar, which is non-contractile, leaving a superficial cicatrix, and it also increases the flow of blood. Its catalytic, destrophic, resorbent, and denutritive effects are more marked than those of the positive. The negative pole should, therefore, be used in all cases of metritis and endometritis when it is not contra-indicated by successive hæmorrhage, leucorrhœal, or hydror-rhœal discharge.

The electric current cauterizes the mucous membrane the same as a chemical cautery. Added to this are the trophic, nutritive, and electrolytic effects, which can be obtained by no other means, and which have long been known to have the double action of destroying abnormal tissue and stimulating new tissue, and hastening the development of a healthy growth at the same time. The result attained depends largely upon the strength of current used, and should be given once or twice a week and from 50 to 200 milliamperes used.

Peri-uterine Inflammations.—Not wishing to go beyond the therapeutical consideration, I will divide peri-uterine inflammation into acute, sub-acute, and chronic, as it matters little whether it be a case of parametritis, perimetritis, or general cellulitis. The treatment of the acute stage is both palliative and curative, for which the faradic battery is used. The difference in the action of the current of a secondary faradic coil, according to the size and length of wire of which it is constructed, should be thoroughly understood (see Chapter I.). This should be constantly borne in mind in treating the acute stage of cellulitis, for the coarse coil, which would be used to cause contraction of a sub-involuted uterus, would aggravate every symptom of the case. The method of application is by the double vaginal electrode, the end of which is placed in as close contact with the inflamed part as possible. The current, which at first should be imperceptible, is gradually increased, but never made strong enough to aggravate the pain. The length of time the application is to continue should vary with the severity of the pain. It should always be given until the pain is relieved, and this may require fifteen minutes to one-half an hour. The period of relief will increase after each treatment until, perhaps, two treatments a day will keep the patient perfectly comfortable. In addition to its sedative effect, it has a very marked action in arresting the inflammation, cutting short the attack, and thus preventing suppuration.

When the acuteness of the disease has subsided so that a sound can be safely introduced into the uterine cavity, the disease may be said to be in a sub-acute stage. The double uterine electrode should now be substituted for the double vaginal electrode and the treatment continued as before, as it is the desire to continue the same effect in the uterus that was produced in the vagina. After the sensitiveness has subsided, the galvanic current should be used, instead of the faradic. The platinum electrode, to which is attached the positive pole of the battery, is introduced into the uterus and treatment given varying from four to eight minutes, with a strength of current varying from twenty to fifty milliamperes.

The negative pole is more efficacious in removing inflammatory deposits. It should, therefore, be substituted for the positive as soon as, in the mind of the operator, the patient is able to bear it. But, as it has a tendency to produce congestion, it should not be used until all the acute congestion is relieved and the sensitiveness reduced to a minimum. This treatment may be continued with the negative inside the uterus until the patient is cured. But a surer and quicker method is to use the galvano-puncture. This, of course, is not done until the disease has become chronic. The number of punctures required varies with the extent of the disease. One puncture may cure a slight parametritis, while three or four may be required to reduce a general cellulitis. The method of operating is as follows: After locating the exact point of inflammation with the finger, the hard rubber shield is placed in position so that its end comes directly against the inflammation. The pointed electrode is now passed through the shield into the substance of the tissue, care being taken to introduce it in a direction so as not to involve the peritonæum or other organs. In a few days contraction will take place, an eschar will separate and a slight exudation appear. The patient should be kept in bed for two or three days, when she will be able to go about without harm. The inactive electrode is placed over the abdomen. With all treatments an injection is given, both before and after, of some antiseptic solution, and, when a puncture is made, the vagina should be plugged with some antiseptic material. When a case has been unfortunate enough to go on to suppuration, we may also use the galvano-puncture to great advantage. When pus has collected in sufficient amount to warrant interference, a puncture is made into the cavity and a very strong current, say 250 milliamperes, should be given for twenty minutes. This will make an opening for the escape of the pus, and also for the introduction of any antiseptic solution that may be desired. No disease is more easily aggravated by rough treatment. It will, therefore, require good judgment and gentle manipulation to secure good results.

Fibroid Tumors.—There are two methods of treating fibroid tumors by electricity. One is by introducing one electrode into the uterine canal ; this is the galvano-chemical caustic method ; and the other is by introducing a needle directly into the substance of the tumor, to which one pole of the battery is attached, or electrolysis.

Galvano-Chemical Cauterization.—The secret of success in treating fibroid tumors with electricity is in using a very strong current. In my experience at least 75 milliamperes must be used to cause reduction. A much weaker current often relieves the subjective symptoms ; but, if possible, from 100 to 250 milliamperes should be used. The clay electrode described in Chapter I. should be placed over the abdomen and connected

Fig. 62.

with one pole of the battery. The internal electrode should be of platinum when it is attached to the positive pole ; but it may be made of other material when attached to the negative. Some use an electrode which is insulated by a rubber or celluloid sheath, which only protects the vagina ; but one so insulated that it protects the internal os is best (Fig. 62), as the pains caused by a very strong current are much less when this part is protected, and we get no benefit from cauterizing it. After the electrodes are placed in position the current should be gradually turned on until it reaches its maximum, where it is continued for from five to ten minutes and then as gradually decreased.

One thing should be constantly kept in mind in treating these cases, and that is if the tumor is hæmorrhagic, the positive pole should be attached to the internal electrode, which should be placed in different positions so that the whole surface of the mucous membrane is thoroughly cauterized by its action. If this cannot all be accomplished at one treatment different portions of the endometrium should be cauterized at different treatments. When it is non-hæmorrhagic, the nega_ tive pole should be used inside the uterus (for reasons concern- ing this the reader is referred to Chapter II.). Treatment may be given for from five to eight minutes, and, if the patient bears it well, may be repeated twice a week.

The Puncture Treatment—Electrolysis.—This method of treatment is more rapid and complete than the galvano-chemi- cal caustic, and should always be tried if that method of treatment fails to produce the desired result. There are two methods of puncturing a fibroid tumor, the internal and the external.

The internal puncture is made through the vagina. The pa- tient is placed on her back in the position for a bivalve speculum. The finger is introduced and a place selected for the puncture, which should always be either through the cervix or in the posterior or lateral cul-de-sac (never in the anterior), and should be the most prominent point, as near the body of the uterus as possible and in a direction corresponding to its axis.

The hard rubber shield is then placed so that the end comes in direct contact with the spot selected; the needle is now passed through the shield into the tumor, but it should never be introduced more than from one-half to one inch.

Some use an uninsulated needle and protect the vagina by a shield the same as the uterine electrode; but an insulated needle (Fig. 63) the same as is used for external puncture, ex_ cept it be much longer, is preferable. This treatment is quite painful and will often require an anæsthetic. Cocaine modifies the pain very much, but it is not so effective here as it is with the external puncture. The best way is to have a long needle attached to a hypodermic syringe; pass the needle through the shield, and inject from ten to fifteen gtt. into the place selected. I first introduce the needle about one-half of an inch, and then gradually withdraw it as it is being emptied. After removing the hypodermic needle, the shield is held carefully in place for four minutes, when the needle used to make the puncture is passed through it into the tumor. An injection of some antiseptic solution should be given both before and after the treatment, and, after the puncture, a strip of antiseptic gauze should be placed in the vagina well up against the cervix. The needle is always attached to the negative pole, and the clay electrode (which is placed the same as with the galvano-chemical cauterization) attached to the positive pole.

The external puncture is used when the tumor is very large and causes a bulging of the abdominal wall, and when it is not convenient to puncture internally as in floating fibroids. Experience teaches that it is just as harmless as the internal method, and is certainly much easier and very much less painful. Great care should be taken in selecting the point to puncture. When a tumor is so large that it extends the abdominal wall, palpation is all that is required; but when it is movable or has become reduced in size, great care should be exercised. In such a case it may be necessary to bandage the upper part of the abdomen, so as to crowd its contents into the lower part, and then the tumor can be felt.

In selecting a place for puncture, care should be taken to

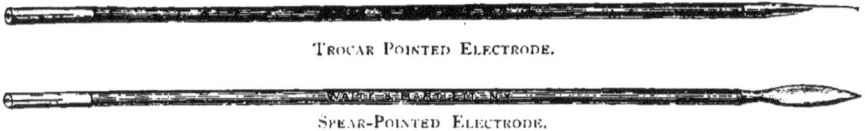

TROCAR POINTED ELECTRODE.

SPEAR-POINTED ELECTRODE.
Fig. 63.

avoid blood-vessels and to choose the most prominent part of the tumor, unless that has been punctured before, when a less prominent one should be chosen. After the operator has determined upon the place, he should next satisfy himself that nothing but the peritoneum intervenes between the abdominal wall and the tumor. A solution of cocaine is then injected into the spot where the puncture is to be made. A 4 per cent. solution is best, and, with an abdominal wall of ordinary thickness, ten drops will suffice to cause complete anæsthesia. In introducing the hypodermic needle it should first be passed down to the peritoneum and gradually withdrawn as its contents are discharged, the larger portion being deposited just underneath the integument. When the abdominal wall is very thick twelve or fifteen drops should be used. A small ring with an opening about one inch in diameter is placed over the site of the puncture and pressed hard against the abdominal wall so as to obstruct the circulation. In about four minutes the cocaine will have produced its maximum of anæsthesia, and the needle (Fig. 63) should then be introduced. No matter from what point this may be, the needle should always point toward the center of the tumor. Care should be taken to introduce it far enough, and not too far, for, if the exposed point of the needle comes too close to the edge of the tumor, it will cause a soreness that will continue for three or four days. The needle is insulated with hard rubber, leaving an exposed surface from one-half to three inches, which is chosen in accordance with the size of the tumor. The strength of current that should be employed varies with the patient, but from 100 to 250 milliamperes should, if possible, be given in all cases, either in puncture or cauterization.

There is one rule which should always be strictly observed in both the external and internal puncture, and that is, never allow the patient to feel the current at the needle. If the cocaine has not produced complete anæsthesia, the patient may feel an ache around the needle, or she may imagine she feels something ; but these pains, either real or imaginary, are not made worse by an increase of the current strength, while, if the pain was produced by the electric current, it would be increased by each additional cell.

The thumb or finger should be pressed firmly over the opening as soon as the needle is withdrawn. This is to prevent air from going into the wound, and also to prevent any hæmorrhage, either external or internal. After two or three minutes all danger of hæmorrhage will be past, and the dressing may be applied. Place a piece of antiseptic gauze over the wound and fasten it with strips of adhesive plaster.

While it is not possible to entirely remove a fibroid tumor by this means, if the treatment is given of sufficient strength and continued long enough, it will almost invariably reduce the tumor more or less, completely relieve the annoying subjective symptoms, the adhesions will be broken up and the tumor will become movable ; the patient will feel better, fat will accumulate in the abdominal wall (a condition which always denotes progress), and the general health of the patient will improve after the shock of the first few treatments disappears.

One class of tumors have not been very successfully reduced either by electro-cauterization or puncture, and that is the fibro-cystic tumors when accompanied by a hydrorrhœal discharge, and when the tumor is very vascular ; but, while these tumors are not often reduced, one very gratifying result is nearly always attained, and that is the improvement in the general condition of the patient. I have lately succeeded in reducing one of these tumors with the positive electro-puncture, using a large platinum needle.

Ovarian Tumors.—The method of treating ovarian tumors is by the external puncture, for the details of which the reader is referred to the treatment of fibroid tumors, the only difference being that the needle is attached to the positive pole instead of the negative. The needle should be made of platinum. Great care will have to be exercised in making the puncture after the tumor has become very small, as it cannot be felt like a fibroid, but should be made out by percussion. As long as the electric current is not felt at the end of the needle, it is safe to suppose that it is in the tumor and no harm is being done. But if, after turning on the current, sensation is felt at the needle, the operation should at once be discontinued. About two treatments a week, varying from 50 to 150 milliamperes in strength and continuing twenty to thirty minutes should be given. This will be found to be much better than at longer intervals. The urine will be greatly increased for from twenty-four to thirty-six hours following the treatment. The patient should be kept in bed for twenty-four hours following the treatment, when, if no soreness exists, she may be allowed to take light exercise.

The success attending this method of treatment varies with different operators. So far its success has been limited to the unilocular variety. The drawback to the operation is not in the power to remove the tumor (for I believe that any unilocular tumor, either ovarian or parovarian, can, by proper treatment, be removed by electricity), but that they often return. Out of five cases of parovarian tumors treated by me in the last three years two returned, and of two cases of ovarian tumor one returned ; but the cases which returned were not treated as thoroughly as the others were. Of course, some may yet return. With the last three cases, however, an iron needle was used at the last treatment instead of the one made of platinum, and a current of sufficient strength and duration given to cause oxidation of the needle, and none of them have since showed signs of return ; but, as they are all recent cases sufficient time has not elapsed to form any judgment as to their cure. I was led to do this by my experience in hydrocele. If a platinum needle was used, the hydrocele would often return ; but I have yet to see one return when an iron needle was used.

[Since this work was placed in the printer's hands the author treated what was diagnosed as a parovarian cyst with an iron needle. One hundred milliamperes were passed for twenty minutes. The operation was followed by soreness and slight fever, which is very unusual, and, instead of the tumor being reduced, which has never failed before with the author, it immediately enlarged, causing great distension of the abdominal wall and discomfort to the patient. On my advice, the tumor was removed ten days later, when it was found to be ovarian and to contain five separate cysts, one of which was filled with pus. Whether the treatment caused this or not is a question ; but it might have done so.]

Ovarian Irritation.—This disease is not, as a rule, very successfully treated by electricity. The method employed is to pass a galvanic or faradic current through the ovary. The large ball electrode, well covered, is pressed up in the vaginal cul-de-sac as near the ovary as possible, and the flexible hand electrode placed on the abdomen over the ovary. If the galvanic current is used, the positive pole should be attached to the ball electrode, and, if the faradic is used, the fine coil should be selected. The author's experience is that the static spark over the region of the ovary is the most successful method of electrical treatment.

Chronic Inflammation of the Ovaries.—The method described above for treating ovarian irritation with the galvanic current (the positive pole attached to a large ball electrode and negative over abdomen) is the best method of treating chronic inflammation of the ovary. Temporary relief may often follow this treatment, but permanent cures are seldom made.

Stenosis of the Cervical Canal.—The best method of treating this disease is to introduce as large a sound as will pass just through the internal os and attach to it the negative pole, while the positive is placed over the abdomen, and give a current as strong as can be borne—even 150 to 200 milliamperes may be employed. One or two such treatments will suffice to cause a cure, or they may be treated on the same principle as stricture of the urethra (which see), and for which the straight Neuman sound is very useful (Fig. 57), but this method of treatment is much slower in its results.

Flexions of the Uterus.—Good results will as a rule follow electrical treatment for flexions of the uterus. A sound so curved that it will pass into the uterine canal should be introduced, and the other placed over the abdomen ; the galvano-faradic current should be employed, using the coarse coil of the faradic battery, unless the parts are very sensitive, when the fine coil may be employed. One treatment may be given a week for a few times, when the flexion will generally disappear. Exactly how this cures the flexion is impossible to say, but it is undoubtedly due to its effect on the muscular coat of the uterus.

Uterine Displacements.—Electricity undoubtedly holds a higher place in the treatment of uterine displacements than any other one agent ; but, in order to be successful, it should be used in connection with some of the other well-known methods which I do not deem it necessary to enumerate. The principle on which electricity is used is to stimulate the muscles and coats of the blood-vessels and thus stimulating the nutrition to the parts. For this the coarse coil of the faradic battery may be used. It is best to use the vagino-abdominal, utero-abdominal and bipolar methods at different treatments, which should be given about three times a week. If adhesions tend to hold the uterus out of place they should be first broken up and absorbed by passing a galvanic current through them.

Goitre.—This disease has been treated by applying the electrodes to the surface of the tumor. Some saturate the sponge that is attached to the positive electrode with iodine and depend on the cataphoric action of the current to carry it through the tissue into the substance of the goitre ; but the best and quickest method is the galvano-puncture. Both poles may be introduced ; but one is preferable. The negative is more energetic in its action, but it also produces more inflammation and is more liable to cause an eschar. The needle should be thoroughly insulated where it comes in contact with the integument, thus preventing electrolysis of the skin and, consequently, avoiding an eschar. It should also be small, the non-insulated part being about one-fourth of an inch in length and inserted about one-half of an inch. The other electrode may be applied to the surface of the back or chest, or held in the hand of the patient. From five to fifteen milliamperes should be given for from ten to fifteen minutes. Two or more punctures may be made at the same sitting, providing the patient can bear it. Two sittings a week may be given, but should not be repeated if any inflammation is present. The success attending this treatment varies with the size and consistency of the tumor.

A small and soft tumor will often entirely disappear. If it is large and soft, the same result may be attained ; but it is not so certain, yet considerable reduction may be expected ; if hard some reduction may be looked for, but if at all large a complete cure is almost an impossibility.

Nævi—Erectile Tumors.—In treating erectile tumors, great judgment will have to be exercised to produce just enough coagulation and not too much. If the electrolytic process is not carried far enough the coagula will be absorbed and circulation re-established. If it is carried too far, sloughing will be the result and an eschar left. Sloughing is particularly liable to occur in poorly nourished children. For the electrolytic action of the two poles when in contact with blood the reader is referred to Chapter II. The best method of treating a nævus is to introduce a needle connected with the positive pole, at the base of the tumor. If the tumor is a large one the needle should be removed and reinserted in another place, and so on until the coagula formed by it covers the whole base of the tumor and completely stops the blood supply. This needle may be of gold or platinum, but an iron one is preferable, as its oxidation helps materially to form the clot. If the tumor is very small, the negative pole may be applied to the surface ; but if it is large enough, it should be attached to a needle which is introduced in such directions and places as to act on all of the superficial parts of the tumor. The current should be strong enough and of sufficient duration to cause thorough coagulation of the tumor, which is made manifest by a hardness of the whole mass.

The circulation should be entirely cut off, for if a little remains it will gradually absorb the clot in the rest of the tumor. If the treatment is successful, the tumor will gradually shrink and finally come off in a hard scab. This will require from ten to twenty days.

Aneurism.—A small superficial aneurism should be treated the same as a nævus, but an internal aneurism will require somewhat different procedure. It will be remembered that, when the two poles of a galvanic battery come in contact with blood, a clot is formed around each ; that around the positive is small, hard, and firm, while that around the negative is larger, softer, and readily broken up. In order to get the best effect, we should introduce both poles in an aneurismal sack. This should be done so that the two will not be more than three-fourths of an inch apart. By so doing the clots around the two needles will join together at their edges. In this way we get the advantage of the large clot of the negative pole, and at the same time it is supported by the firm clot of the positive.

The method of operating is to introduce needles into the sack three-fourths of an inch apart until the sack is filled, and then attach every other one to the positive and the others to the negative pole. By this means we get the hard, firm clot of the positive on each side of a soft negative clot. Treatment should be continued until all signs of circulation in the aneurism have ceased and the tumor has become hard and firm.

The needles used should be thoroughly insulated with hard rubber, so as to protect the walls of the aneurism from the electrolytic action, for accidents have occurred from rupture of the wall of the sack when this precaution was not taken. Steel needles are, of course, used on the negative pole, but gold or platinum needles should be used on the positive (I am well aware that the clot would be materially larger if an iron needle were used here, but the difficulty of its removal makes it liable to detach the clot). The success attending this method of treating an aneurism is very good in the hands of a careful operator who understands electro-physics and physiology; but is unsuccessful in the hands of the inexperienced operator.

Cystic Tumors.—Cystic tumors as a rule are successfully treated with electricity. In order to get at a perfect understanding of the manner of treating them, we should divide them into two classes ; those where the membrane covering the sack is their only covering, such as a ranula of the mouth, and those where the integument or other tissues is the outer covering.

In the first class of these tumors the negative pole should be used and the needle not insulated, so as to produce electrolysis of the sack, thus making a permanent opening for drainage. In the second class the positive pole should be attached to the needle, which should be thoroughly insulated. A gold or platinum needle may be used ; but in that case the tumor is liable to return. Therefore an iron one is preferable, and a current of sufficient strength and duration should be given to cause oxidation and disintegration of it. In hydrocele I often try to dissolve part of the needle in the sack. A very large pad should be placed on the surface of the body as near the tumor as convenient, and as strong a current as can be borne given for from twenty-five to forty minutes. One treatment may suffice to cause a cure, but in very large tumors more may be required. Of course those that are treated with the negative pole and a permanent aperture made in the sack disappear immediately, as their contents run out of the opening; but those belonging to the second class will disappear slowly, as their contents are absorbed and thrown off by the kidneys, which is made manifest by the increase in the amount of urine following the treatment.

Removal of Superfluous Hairs.—This is done by a simple process of electrolysis. A very fine needle, connected with the negative pole of the battery, is introduced into the hair follicle, while the positive is held in the hand of the patient, and a current ranging from two to four milliamperes is given for from one-half to one minute, at the end of which time the hair can be pulled out of the follicle with perfect ease, which is proof that it is dead and will not return. If the hair sticks the needle should be reintroduced and the current allowed to pass again. An ordinary depilatory applied to the surface will kill the hair, so that it can be easily removed, but it will immediately return; therefore it is evident that the process of electrolysis does more than to just loosen the hair. According to our present knowledge of the function and anatomy of the hair follicles, there are three layers of cells lining it. The first layer furnishes the nutrition, the second the epithelium for the hair, and the third the pigment or coloring matter. It is evident from this that, in order to prevent a return of the hair, the first two layers of the follicle must be destroyed. This is what electrolysis does. It is also evident that the needle must be introduced into the sack so as to act on all sides of the follicle, and great care should be taken to pass it to the bottom of the sack and not through its sides. To do this will require a good light, a steady hand, and a good sight. One who has had much experience will also rely as much on the sense of touch as on his sight. If the needle is at the entrance of the sack it will go in without force; but if it is not, some slight pressure will have to be made. The depth the needle should be introduced is gauged by the length of the hair-root, generally from one to two millimetres. During the passage of the current some slight froth will accumulate around the needle.

Some use a handle containing an interruptor, so that the circuit is not closed until the needle is in the sack. This always causes a shock, and is very much more disagreeable than the solid handle where the current is passing continually. Some soreness will remain after the operation, and if the treatment has been severe and the hairs taken off close together (which should never be done), there will be quite an exudation.

The mode of treatment after the operation may be varied according to the fancy of the operator. The author puts on a little vaseline and then powders the face thoroughly with boracic acid. Just enough vaseline should be used to make the powder stick to the face.

Ulcers.—In chronic ulcers electricity may be of service in promoting healthy granulations. The best method is to use a bulb electrode attached to the negative pole of a galvanic battery, and go thoroughly over the whole surface of the ulcer, so as to obtain the galvano-chemical-caustic action of the current. From five to ten such treatments should be given, and iodoform or some other antiseptic dressing applied afterward and await results. In about a week healthy granules will appear. These will spread over the whole surface and the ulcer gradually heal. If certain spots that are not healthy make their appearance they should be given a second treatment immediately, or they will have a bad influence on the healthy portion.

If the ulcer is particularly hæmorrhagic or has a profuse discharge, the positive pole may be attached to the active electrode, in which case the bulb should be made of platinum.

Deformities.—After operations for deformity, electricity will be of great service in restoring the weak and atrophic muscles that are the result of the deformity. The anode should be placed over the spinal origin of the nerve which supplies the affected muscles, and with the cathode give interrupted applications several times over all the motor points.

In all that class of diseases which come under the orthopædic surgeon's hands electricity is also of the greatest service. Treatment should be given to the weak muscles the same as described above.

General electrolization, as explained in General Therapeutics, may be given, causing several contractures of those muscles that are most affected. The latissimus dorsi should be vigorously contracted in lateral curvature of the spine. When the deformity is due to spinal disease the current should be given through the diseased portion, to improve the nutrition in those parts.

Boils and *abscesses* may sometimes be arrested if taken before the suppuration begins, by applying the anode stable over them. The stable application of the cathode may be used to hasten suppuration when the disease has progressed too far to be stopped.

Cancer.—Many cases of cancer have been reported cured by electrolysis, but after quite an extended experience in the treatment of cancers, the author has come to the conclusion that electricity is one of the very worst applications that can be made for this disease. After electrolysis has been performed in a cancer it begins to break down around the puncture, a fungus is liable to form over it, sloughs are thrown off, and it is almost impossible to heal it up. At the same time the patient's general health begins to fail, and it seems to fail just in proportion to the sloughing. In fact, every symptom, both local and general, is aggravated by the treatment. The method recommended by Rockwell, to perform electrolysis of the edge of the wound after the cancer is cut out, invariably prevents healing by first intention and greatly retards healing by granulation; and I have known the cancer to return before the wound had entirely healed in amputation of the breast for this disease, when this method was employed. The author feels that electrolysis used in any form of a cancerous condition cannot be too strongly condemned.

Stricture of the Œsophagus.—This disease is treated on the same principle as stricture of the urethra. The size of the opening through the stricture should be first ascertained by means of œsophageal bougies, an electrode with a bulb from one to two centimetres larger than the stricture introduced, and a current of from five to ten milliamperes given until the instrument passes through the stricture without pressure. The operation in very urgent cases may be repeated during one sitting, until a sufficient opening has been attained to inject milk or other liquid food into the stomach. The treatment should then be repeated about every ten days until the operator thinks the opening is of a sufficient size.

Enlarged Glands.—Enlarged glands may be treated by passing a galvanic current through them, or by electrolysis. When treated by the former method, one electrode is pressed firmly over the enlarged gland, and the other, a large electrode, is so placed that the current passes directly through the gland. If the gland is particularly sore and tender, the positive pole should be placed over it; but if it is not, the negative should be held on the gland, although it is better to get the action of both poles alternately. Treatments may be given every other day. About 40 per cent. of the cases thus treated have been cured. A quicker and surer method is to introduce an insulated needle and produce electrolysis. The negative pole should be attached to the needle and the positive applied to the surface of the body. A current of from twenty-five to fifty milliamperes should be used and continued for from ten to fifteen minutes. The operation may be repeated as soon as the inflammation following the preceding one has subsided. It will generally require from five to ten such treatments to cause complete reduction. About 70 per cent. have been cured by this method.

Ankylosis.—Where this disease is due to inflammatory exudation and is not osseous, good results may be expected from the use of galvanism. The current should first be passed in various directions through the affected point to obtain the catalytic effect, and thus cause absorption of the exudation. Next the muscles should be thoroughly contracted by applying the current to the motor points.

CHAPTER VII.

GALVANO-CAUTERY.

FOR the principles underlying and the construction of a cautery battery the reader is referred to Chapter I. The galvano-cautery is used for the same purpose as any other cautery, for the battery acts only as a heat producer.

Both knifes and loops are used in operations with the gal-

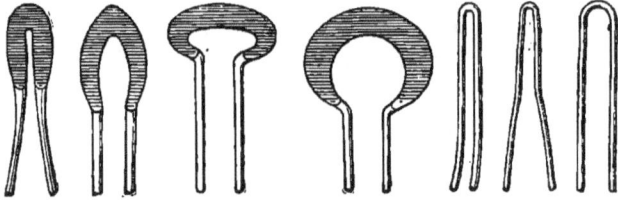

Fig. 64.

vano-cautery. A knife is a very small loop fastened to the end of two copper wires (as in Fig. 64). Cautery knives may be made in the various shapes and styles required to suit particular

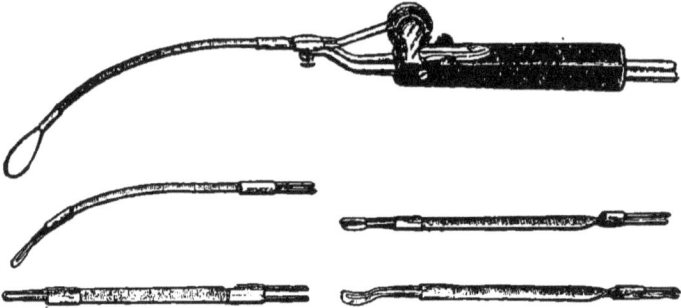

Fig. 65.

places to be operated upon. A loop is a platinum wire so placed in a handle as to form an écraseur (Fig. 65).

The galvano-cautery may be used for any purpose for which a cautery is required, but only those in which it is particularly indicated over any other form of cautery will be considered here. If an acid battery is to be used for an operation that will require twenty or thirty minutes, it should be filled with fresh fluid. If a storage battery is to be employed it should be thoroughly charged before beginning the operation, and the battery should be tested so as to see that it is in order, for the electro-motive force of a cautery battery is so low that dirt between the connectors or anything that tends to cause resistance may entirely destroy its action.

Operations with the Cautery Ecraseur—Amputation of the Tongue.—Under this head are included all operations on the tongue where a cautery loop is used, whether it be a complete amputation or whether a piece is to be taken out of the side of the tongue, as the principle is the same.

After the patient is etherized, the mouth is held open by Whitehead's gag (Fig. 66), a strong ligature is passed through

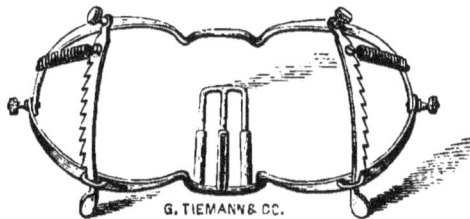

G. TIEMANN & CO.

Fig. 66.

the healthy portion of the tongue and its ends tied to form a loop to draw the tongue out with, so that organ can be seen. The parts to be removed are then thoroughly isolated by running pins through the tongue just inside the diseased tissues. The ordinary surgical pins are too long for this purpose and are exceedingly difficult to introduce. A short, round surgical needle is best, and should be introduced with a pair of large needle forceps. The needles should be introduced about one-fourth of an inch apart. The platinum wire loop is then placed behind the needles and the écraseur tightened.

The success of the operation, so far as its being bloodless is concerned, depends upon the remaining part of the operation. Mr. Bryant, of England, has for many years been the most prominent authority on galvano-cautery operations on the tongue. In a lecture, published in the *Lancet* of February 28, 1874, after giving the primary details of the operation, he says: "It (the wire) should not be heated beyond a red heat, and the redness ought to be of a dull kind. Above all, the process of tightening should be very slowly performed, the wire of the écraseur being screwed home only as it becomes loose by cutting through the tissues." The lecture from which the above was taken is quoted either in part or wholly in nearly every text-book on electro-surgery, and in a great many works on general surgery ; consequently it has become the most popular method of operating on the tongue. It is to the last sentence of the quotation given above that attention is called, for in it lies the cause of Mr. Bryant's operation (which in all other respects is so perfect) being a failure, so far as the blood-

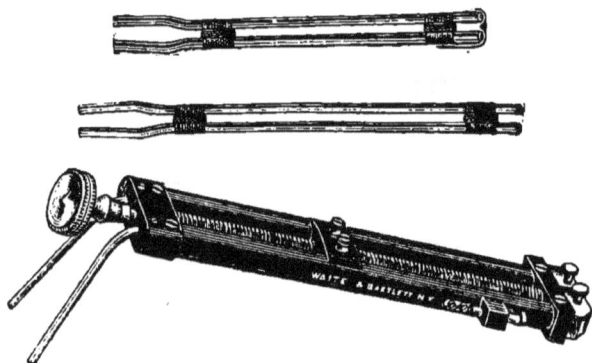

Fig. 67.

less operation is concerned. A large platinum wire, No. 22 of Brown and Sharp's gauge, should be used. This is important ; for a smaller one may break and thus delay an operation. The handle (Fig. 67) should be composed of two strips of hard rubber tubing placed parallel, separated from each other about one-half an inch and held in position by two blocks of hard rubber, one at each end. Running through the rubber tubing are two large copper wires, one of which is cut with a connector so placed that the operator can open or close the circuit with the thumb of the hand in which he holds the instrument. Between the hard rubber tubes is a screw running their whole length and which is held in position by the blocks. This screw has a large milled wheel which gives the operator a powerful traction. Sliding on the screw is a hard rubber block with strong pegs for fastening the loop of the écraseur to. This block is first screwed down to the end of the instrument nearest the loop, and the wire is placed in position and fastened to the pegs. The screw is then turned until the wire is imbedded in the tissues of the tongue ; the wire is allowed to heat just enough to cut its way with constant traction on the loop. To do this the wire need never come to a perceptible redness. The operator should keep constant and strong traction on the loop ; in fact, use it as you would an ordinary cold écraseur, except that the traction used should require a very small degree of heat to complete its work. It is this one point, the constant and strong traction, that should be changed in Mr. Bryant's operation. For it is on this that the success of the operation, so far as its bloodless character is concerned, depends. The advantages derived from this method of operating over that of Mr. Bryant's are :

First. The tissues are rendered so tense by the traction that much less heat will sever them than when the wire is left loose.

Second. You will be able to maintain a much steadier heat, as the wire is constantly imbedded in the tissues ; while with the Bryant operation the wire cuts itself loose, the heat rises rapidly, and, if the operator is not continually on his guard, it will rise to a white heat before he is aware of it.

Third. You get all the combined advantages of the cold écraseur and the cautery. The walls of the blood-vessels are drawn tightly together, little clots form in them, and the glutino-fibrous exudation around the wire glues, as it were, the closed ends of these vessels together ; while, at the same time, you also get the advantage of the cautery. With the Bryant operation you get none of these effects. To repeat his own expression, "the wire of the écraseur being screwed home only as it becomes loose by cutting through the tissues." It is evident that in this case the ends of the vessels are not drawn together and no clots form in them. If, when the wire is cutting itself loose, it should cut into the side of a vessel that is not constricted and with blood coursing through it, the blood will be sure to escape by the side of the wire ; when, if the vessel was first constricted and circulation stopped, blood would not flow.

Fourth. This operation can be safely performed with ether, as the wire is continually imbedded in the tissues and need never come to a red heat, consequently the vapor of ether can not ignite.

This operation is generally used for cancerous growths.

Amputation of the Penis.—The wire of the écraseur is placed

Fig. 68.

around the penis at the point where it is to be amputated, and the same details of the operation followed as are given for amputation of the tongue. This operation is generally performed for cases of cancer and gangrene following traumatism or diseases of a specific nature.

Amputation of the Cervix and Removal of Uterine Polypus.—This is one of the most useful fields of operating with the

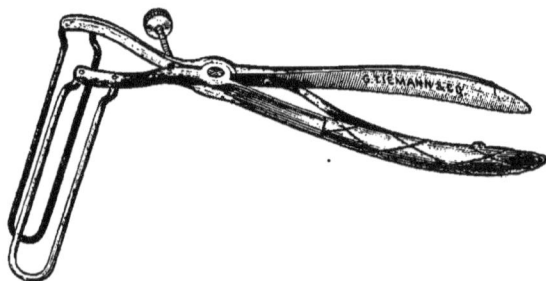

Fig. 69.

galvano-cautery. The loop is passed up around the neck of the polypus or around the cervix, if that organ is to be removed, and the same details followed as are given under amputation of the tongue.

Nasal polypi are also removed in the same manner, but require a smaller handle and loop, as in Fig. 65.

Operations with the Galvano-Cautery Knife.—As has been stated, the galvano-cautery knife may be used in any place where an actual cautery is indicated, but it is particularly useful in small apertures such as in the nasal cavity or throat.

Hypertrophy of the mucous membrane over the turbinated bones may be very effectually removed by the galvano-cautery knife. A handle (Fig. 68) containing an interruptor, so that the heat is not turned on until the knife is in position, should be selected. The nostrils are held open by means of a self-retaining speculum, so as to give the operator the use of both his hands. A small pledget of absorbent cotton, saturated with a 4 per cent. solution of cocaine, is first applied to the surface to be treated and left in that position for from three to five minutes, when it may be removed and the cautery applied. Only the dullest red heat should be used, and the cauterization should be superficial only. The second treatment should not be given until two weeks have elapsed, and not then if the previous cauterization is not completely healed. As there is contraction after each cauterization, great care should be exercised not to cause too much destruction of the tissues, or atrophy will be the result. The number of operations required will vary with the amount of hypertrophy. Generally from three to four will suffice.

The galvano-cautery is also used to reduce the size of enlarged tonsils. A pointed knife is used to puncture the crypts of the tonsil. The knife is first introduced and then slightly heated for about five seconds. Four or five crypts may be punctured at one sitting, and the treatment may be repeated as soon as the soreness and inflammation (which are never great) have subsided. The tonsils gradually shrink down to their normal size.

Hæmorrhoids may be removed with the galvano-cautery loop. The rectal speculum (Fig. 69) is introduced, the hæmorrhoids drawn down, and the loop passed around each close to the body and gradually severed with a dull red heat. This is a very successful method of treating this disease, and only slight soreness follows the operation.

INDEX.

151

9 783337 778880